HOW TO DISSECT

EXPLORING WITH PROBE AND SCALPEL

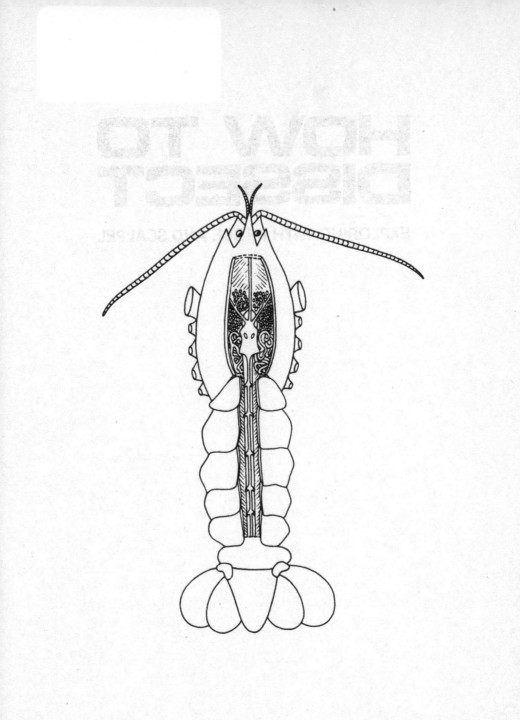

HOW TO DISSECT

EXPLORING WITH PROBE AND SCALPEL

Fourth Edition

WILLIAM BERMAN

Chairman (Retired) of the Department of
Biological and Physical Sciences
S.J. Tilden High School
Brooklyn, New York

A FIRESIDE BOOK
Published by Simon & Schuster
New York London Toronto Sydney

F

FIRESIDE
Rockefeller Center
1230 Avenue of the Americas
New York, New York 10020

Published in 1986 by Prentice Hall Press
Previously published by Arco Publishing, Inc.
Drawings prepared from author's sketches.

FIRESIDE and colophon are registered
trademarks of Simon & Schuster Inc.
First Fireside Edition 1992

Manufactured in the United States of America

30 29 28 27 26

Library of Congress Cataloging-in-Publication Data

Berman, William.
How to dissect.

Includes index.
Summary: A guide for dissecting animals, beginning with
the earthworm and progressing to more complex anatomies
such as grasshopper, starfish, perch, and ultimately a fetal
pig. Includes a chapter on dissecting flowers.
1. Dissection—Juvenile literature   [1. Dissection.
2. Anatomy]   I. Title.
QL812.5.B4   1984   591.4'07'8   83-27510
ISBN-13: 978-0-671-76342-8
ISBN-10:   0-671-76342-3

Contents

of shark. Digestive system. How to expose heart and main veins. The hepatic portal system. Main arteries. Reproductive organs of male and female. Nervous system.

Special Projects: Comparative study of hearts—sharks and other specimens. Tips from scales—identifying species from their scales. Evolution in the aquarium.

Acknowledgments

To my wife, Betty, whose skillful editorial incisions helped to shape this book, and to the publishers, whose insatiable "whys" brought the author closer to the reader.

Preface to the Fourth Edition

HOW TO DISSECT ORIGINALLY WAS INTENDED TO BE A CLEAR AND uncomplicated guide for students with little or no experience in dissection. Another objective was to articulate the dissections to produce the effect of an interconnected series paralleling the biological emergence of the groups represented by the individual dissections. Finally, the book was designed to humanize the disparate dissections by revealing anatomical relationships that ultimately result in a better understanding of our place in the world of living things.

As the writing progressed, the need to expand the scientific literacy of the dissectors became more pressing. The earlier chapters were written with simplified text and directions. Gradually, the experience gained in each chapter became a ladder of incremental learning for the next dissection, until the level of the final chapters was considerably heightened. This basic plan remained a self-injunction to balance simplicity with the need for sophistication, rather like moving from freshman to sophomore and toward the senior status for the two new chapters, "The Clam" and "The Perch."

The clam was included in this new edition to provide a better appreciation of a very large taxonomic group of animals common to human experience. The perch was chosen as an example of the group of animals that dominates the waters of the world, a group whose diverse development included organisms that foretold a future of the subphylum Vertebrata.

Somewhat more emphasis has been placed on the rationale of taxonomy and on the probable lines of evolutionary development. A taxonomic summary of chordate classification and an illustration of the Evolutionary Tree of Life that can be useful throughout the book are included at the beginning of the chapter on the fetal pig.

Twenty-two new illustrations and several new projects have been added. The book now covers the spectrum of most organisms generally used for dissections in science classrooms and laboratories and should have greater value for students and their instructors.

Once again I express my admiration and respect for Betty, my wife, companion, and personal editor, for her devoted assistance and time in preparing the fourth edition.

William Berman
1984

HOW TO DISSECT

EXPLORING WITH PROBE AND SCALPEL

1

DISSECTION IS ADVENTURE

YOU ARE ABOUT TO EMBARK ON A NEW KIND OF ADVENTURE. WITH probe and scalpel you will explore the anatomy of animal and plant life. Dissection not only reveals the architectural plan of living creatures; it also shows how life has evolved from the simple to the most complex forms.

This is not a "cookbook" of dissection "recipes." Our directions for dissecting must be, and we hope they are, clear and easy to follow. But you should go beyond the limits of dissection itself. For the inquisitive young scientist with questions which exceed the scope of this book, we have included suggestions for more advanced work. For example, in addition to dissecting the earthworm, you might want to study its ability to solve problems! In addition to dissecting a flower, you may want to look into the problem of developing a method of supplying food and oxygen for our travelers in outer space. Dissection is a tool, a technique of investigation used in the service of science.

You will retrace the steps of some of the great biologists. You will travel along roads branching off to other roads, and at the end you may find yourself in new bypaths, yet uncharted by science. In these days of expanding scientific knowledge, young scientists often find themselves advancing toward the frontiers of research. You will meet some of the unsolved problems and stirring challenges facing our scientists. Perhaps some day you will help find the answers to some of these problems.

In your work with dissection you may find the answer to your own future. Do you want to be a doctor, a science instructor, a dentist? Perhaps you would rather be a chemist or a physicist, doing research in cancer, in heredity, and in many other vital areas. The choices are many and are often bewildering. Few of us make early decisions. We all need guidance. This book can help give you a start in mastering background you will need before you can determine whether you are suited for a profession in science. Besides promoting

1

a knowledge of anatomy that will increase your understanding of the human body, it can provide evidence of your own aptitudes and capabilities. You may decide to join the dedicated army of scientists and make research your life work.

This book should be especially helpful to biology students from the secondary school through junior college and freshman college levels. It will be useful to students in advanced courses in high school zoology and in advanced college placement biology courses in high school. College students will find that our simplified dissecting techniques and directions will enable them to benefit more thoroughly from their required readings in physiology, evolution, and other related topics.

Most people have a strange notion about dissecting. They see themselves standing in a pool of blood, gripping bloodied instruments. This notion is all wrong. A dissection is clean. The only "blood" present in the prepared specimen is the latex injected into the blood vessels in the biological supply house by which the arteries are stained red and the veins blue. This "color map" helps us trace the circulatory systems. When you receive the specimen it has already been treated with preservatives to keep the tissues from hardening and drying out. All you need do before dissecting is to wash away thoroughly the excess preservative under running water.

Another false notion about dissecting is that all you do is cut and slice. A specimen is not a loaf of bread; it is a marvelously assembled and intricate set of structures held together by tissue, mostly connective tissue. You might compare a specimen to a carton of fragile, expensive dishes you are about to open. Each article is separately wrapped and you are going to unwrap it very cautiously. That's what we do when we dissect; we make careful incisions to expose parts. We then use a probe (often called a seeker), a long, thin, pliable metal rod with a smooth, rounded tip, to separate organs from their coverings. In a sense, we are carefully unwrapping the parts of the specimen without injuring any of the parts. Except for major incisions, *don't cut—dissect!*

All you need to profit from this book is the curiosity you were born with, and the ability to read well enough to understand and to follow directions. However, this is not the kind of book that can be merely read as you would read a novel. Except for the introductory remarks in each chapter, you must work with the specimen before you because the text and the illustrations or diagrams refer to the specimen. It often happens that many organs which may be difficult to visualize from the text can readily be seen in the specimen. The diagrams have been simplified to make it easy to identify parts on the specimen. A picture may be worth a thousand words, but the real

2

thing is better than a thousand pictures. To avoid confusion, very small blood vessels and nerves are not shown in the diagrams. After you have mastered the main techniques of dissection you might consult a more advanced book on comparative anatomy for more technical study.

The dissections have been arranged so that we begin with relatively simple, primitive specimens and work our way up to the more advanced forms of life. We begin with the earthworm, a spineless creature, and close with the fetal pig, a vertebrate animal with a well-developed spine and which, in many structural features, is similar to man. The dissections provide a fascinating glimpse into the story of evolution. The basic techniques for dissecting the earthworm are used in dissecting more complicated specimens. However, each new dissection will require additional skills. When you have dissected the crayfish you will gain the basic training for dissecting the grasshopper. Dissecting the shark is excellent preparation for dissecting the frog, and the frog dissection is a fruitful basis for the dissection of a mammal, the fetal pig. The more dissecting you do, the more you will get out of this book.

IMPORTANT POINTS FOR SUCCESSFUL DISSECTIONS

Before dissecting, read directions carefully and examine accompanying diagrams thoroughly. When dissecting, turn the dissecting pan to the position most comfortable for you as you handle the instruments. Use the probe often for separating structures (veins, nerves, organs) from connective tissue, and for tracing the course of hollow structures by inserting the probe to see where they lead or originate. Trace and master one system at a time. Then see how the different systems are related to each other and to the general body plan.

Dissection is analytical. Separating and analyzing associated parts provides us with the basis for productive thinking when we assemble data to produce new ideas. When you dissect or take apart an organism, you are taking the first step toward putting together, or synthesizing, new theories and new knowledge. Do not be satisfied with dissecting only one example of a whole group of animals. Use the knowledge and skills you have gained in independent learning.

At the end of each chapter you will find ideas for exciting research, for science projects, and for enjoyable hobbies. Make your own diagrams and keep records of your dissections and findings. Compare the systems of every animal and plant you dissect to discover evolutionary connections. Take photographs of the dissections.

Each animal has its own story to tell. No dissection can be adequately meaningful if it is done as a single experience. It is like trying to judge a painting without having examined and analyzed other paintings.

What equipment will you need for dissection, in addition to the specimens? Certain basic tools will be useful in all dissections (see Fig. 1). Individual instruments or fully equipped dissecting kits may be obtained from biological supply houses and hobby shops. Additional instruments with special advantages for particular dissections are described in the text where they are needed. Following is a list of basic equipment:

Dissecting pan
Dissecting pins
Dissecting needles
Specimen jars
A thin probe
Scalpel, medium size
Paper toweling
Single-edge razor blade
Straight-tipped forceps, medium size
Curved-tipped forceps
Hand lens, 5× to 10× magnification
Dissecting microscope, 5× to 10×
Dissecting scissors, 4 to 5 inches
(10 to 13 centimeters, or cm) long

Formaldehyde preserves specimens or organs for future study. A good mixture for preserving most specimens is a 4% solution of formalin, which is made by adding 4 milliliters (ml) of commercial formalin to 96 ml of water. The mixture is a very weak solution of formaldehyde. A little cold cream rubbed into your fingers will prevent the preservative from dehydrating your skin.

Order double-injected specimens by their scientific names as indicated in the text. Double-injected specimens are injected with two different colors—red for arteries, blue for veins.

Several scientific supply houses currently offer specimens for dissection, preserved with chemicals that do not have the irritating and potentially toxic effects of formaldehyde, although some formaldehyde may be used in conjunction with the nontoxic preservatives. The specimens may be purchased in sets, some of which include most organisms generally dissected in schools and universities.

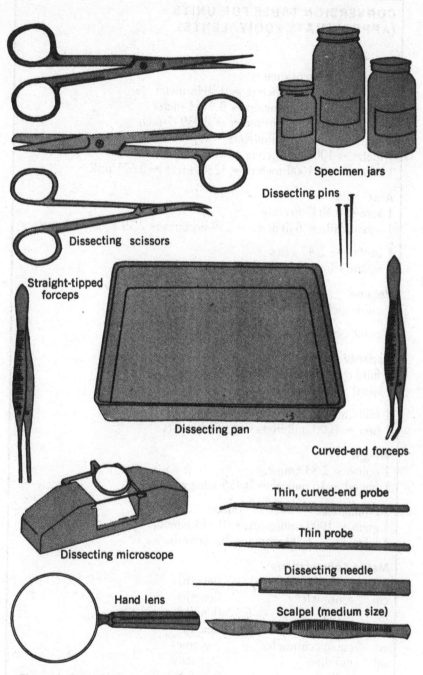

Specimen jars

Dissecting pins

Dissecting scissors

Straight-tipped forceps

Dissecting pan

Curved-end forceps

Thin, curved-end probe

Thin probe

Dissecting microscope

Dissecting needle

Hand lens

Scalpel (medium size)

Figure 1. Basic Dissection Equipment.

CONVERSION TABLE FOR UNITS
(APPROXIMATE EQUIVALENTS)

Length
1 inch = 2.54 centimeters
1 foot = 30.48 centimeters = 0.3048 meter
1 yard = 91.44 centimeters = 0.9144 meter
1 millimeter = 0.1 centimeter = 0.03937 inch
1 centimeter = 10 millimeters = 0.3937 inch
1 meter = 100 centimeters = 39.37 inches
1 kilometer = 1000 meters = 3280.8 feet = 0.621 mile

Area
1 acre = 0.4047 hectare
1 square mile = 640 acres = 259 hectares = 2.59 square kilometers

1 hectare = 2.47 acres
1 square kilometer = 0.3861 square mile

Volume
1 cubic inch = 16.387 cubic centimeters

1 cubic centimeter = 0.06102 cubic inch

Capacity
1 fluid ounce = 29.573 milliliters
1 quart = 16 fluid ounces = 0.946 liter

1 milliliter = 0.0338 fluid ounce
1 liter = 1000 milliliters = 1.0567 quarts

Weight
1 ounce = 2.83 grams
1 pound = 16 ounces = 0.455 kilogram

10 milligrams = 0.000353 ounce
1 gram = 1000 milligrams = 0.353 ounce
1 kilogram = 1000 grams = 2.2 pounds

Metric Abbreviations
mm	millimeter	(length)
cm	centimeter	(length)
m	meter	(length)
km	kilometer	(length)
cc	cubic centimeter	(volume)
ml	milliliter	(capacity)
mg	milligram	(weight)
g	gram	(weight)
kg	kilogram	(weight)

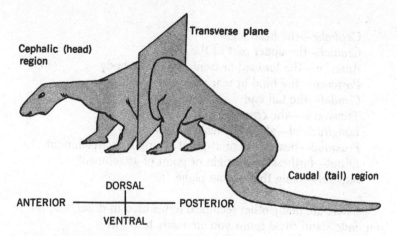

Cephalic (head) region

Transverse plane

Caudal (tail) region

DORSAL

ANTERIOR ———————|——————— POSTERIOR

VENTRAL

The compass points of anatomy (anterior, posterior, dorsal and ventral) appear in all forms of animal life from the simple, spineless creatures to man himself. These compass points help us observe how animal life evolved from a crawling stage to an upright position.

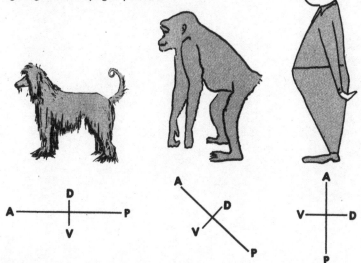

A ——————|—————— P
D
V

A
|
V — D
P

A
|
V —|— D
P

Figure 2. The Compass Points of Anatomy.

IMPORTANT TECHNICAL TERMS

And finally, here are a few of the important technical terms frequently used in dissection (see Fig. 2):

> *Dorsal*—the back or upper part of the animal
> *Ventral*—the abdominal side or lower part of the animal

7

Cephalic—the head region
Cranial—the upper part of the head
Anterior—the forward or front end of the body
Posterior—the hind or rear part of the body
Caudal—the tail end
Transverse—the cross section
Longitudinal—along the length of the body
Proximal—nearest to, or at point of origin or attachment
Distal—farthest from origin or point of attachment
Sagittal—along the median plane

There are many other technical terms used in dissection, but if you understand these terms you are ready to begin.

Did we say begin? Just so!

2

Night Crawlers in the Laboratory

THE EARTHWORM

MANY OF US HAVE HAD THE FUN AND EXCITEMENT OF FISHING WITH an earthworm as bait. But few people realize how important the earthworm is to civilization, for this lowly creature is one of nature's greatest cultivators. Charles Darwin, one of the world's famous scientists, estimated that England's farmland had over 50,000 earthworms to the acre. He calculated that these worms turn over 18 tons (about 16,000 kilograms, or kg) of soil per acre and bring 1 inch (2.54 cm) of rich soil to the surface every five years. While this estimate may not be accurate for all tillable areas in the world, it shows that our little friend the earthworm enriches farmland and helps indirectly to provide more food for a rapidly expanding world population.

Another benefit plants and animals derive from the earthworm is that the soil becomes more porous and air circulates more readily through the earth because of all the burrows or holes made by the earthworm. This helps to support living things in and on the soil.

NIGHT LIFE OF THE EARTHWORM

The earthworm hunts for food at night. That is why it is called a "night crawler." Its food consists of fallen leaves and animal debris. It usually extends its body from a small burrow which it creates by literally eating its way through the soil. The hind part of the worm's body remains near the surface end of the burrow while the rest of the animal forages for food. If the worm is disturbed by its mortal enemy, the robin, who hunts the earthworm during the day, or by any unwelcome nocturnal intruder, it retracts its muscles with remarkable speed and tries to escape into its underground retreat. If the earthworm is undisturbed, it will swallow leaves, small particles of earth, and other material. The worm's digestive system then acts upon this food, with the help of the earth it swallows. In this way the worm gets its nourishment, while its wastes help to fertilize the soil.

Why do we select the earthworm as our first subject for dissection? The earthworm is a lower invertebrate (an animal without a backbone) that has a simplified pattern of structure. A study of the earthworm's anatomy will therefore provide us with a key for unlocking the secrets of structure in higher forms of invertebrate animal life and also will furnish some clues to understanding human anatomy.

MATERIALS NEEDED

For our first dissection we will need preserved injected specimens of earthworms, *Lumbricus terrestris*. More than one specimen should be purchased because the first specimen may be damaged through inexperience. An injected specimen is preferable because the structural systems of the worm, particularly the circulatory system, will show up more clearly. We will also require the following equipment:

- A dissecting pan.
- A quart-size bottle of formaldehyde, to be used as a preservative if the dissection is carried over to another day. (Keep the bottle tightly sealed when not in use to prevent the odor from spreading.)
- Two or three small specimen jars with covers, in which to store the specimens. Specimens may be kept almost indefinitely in formaldehyde.
- Dissecting pins to pin the specimen to the dissecting pan. Keep a good supply on hand.
- A 5× to 10× hand lens to examine small parts of the specimen.
- An inexpensive 5× to 10× dissecting microscope.
- A pair of fine surgical scissors 4 to 5 inches (10 to 13 cm) long with one sharp tip and one blunt, rounded tip for use in cutting through parts of the body.
- A medium-sized scalpel to cut through skin and tough tissues.
- A sharp, single-edge razor blade for starting incisions.
- A pair of curved-end, fine-tipped forceps for use in holding tissues that are to be dissected or examined.
- A thin probe for exploring body tubes and for separating structures from their membranes.
- Dissecting needles for separating organs from their membranes, pinning down dissected parts, etc.

• Paper toweling to clean and dry instruments and to dispose of the dissected specimen in a rubbish barrel or incinerator.

EXTERNAL ANATOMY

Before we begin actual dissection we should become familiar with the external anatomy of the earthworm. With Fig. 3 as a guide let us proceed to examine the specimen. Note that the dorsal (upper) side has a darker coloration than the ventral (lower) side. The bristles (*setae*), which enable the earthworm to move about and hold firm to the ground, are found only on the ventral (lower) side, as are the tiny pores on the body wall. The earthworm breathes through its moist skin. The pores help to keep the skin moist. They connect with the *nephridia*, which excrete liquid wastes collected in the *coelomic cavity* (see Fig. 13 later in this chapter). The pores can be seen with the aid of a hand lens or with a dissecting microscope. Nearer to the anterior (front) end than to the posterior (back) end of the worm is a thick cylindrical collar (*clitellum*) used in reproduction. The worm's protective coating (*cuticle*), a secretion of skin, is kept moist by special mucus glands. Each segment (*somite*) is partially separated from its neighbor by a thin wall called the *septum*. The fleshy lobe (*prostomium*) over the mouth is not considered a segment of the earthworm. It serves to give a cushioning effect to the sensory endings of the earthworm.

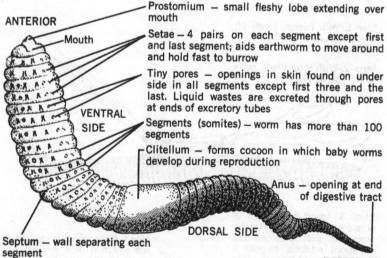

ANTERIOR

Mouth

VENTRAL SIDE

Prostomium — small fleshy lobe extending over mouth

Setae — 4 pairs on each segment except first and last segment; aids earthworm to move around and hold fast to burrow

Tiny pores — openings in skin found on under side in all segments except first three and the last. Liquid wastes are excreted through pores at ends of excretory tubes

Segments (somites) — worm has more than 100 segments

Clitellum — forms cocoon in which baby worms develop during reproduction

Anus — opening at end of digestive tract

DORSAL SIDE

Septum — wall separating each segment

POSTERIOR

Figure 3. External Anatomy of the Earthworm, Turned to Show Ventral Side at Anterior (Front) End and Dorsal Side at Posterior (Rear) End.

ARRANGING THE SPECIMEN FOR DISSECTION

Now we are ready to dissect and study the internal anatomy of the earthworm. Place the specimen on the dissecting pan with its ventral side down. Extend the specimen to form a straight line. Put a dissecting pin through the prostomium and another through the *anal segment* (Fig. 4). To dissect the worm, see Figs. 5–9.

Dissecting pin through prostomium

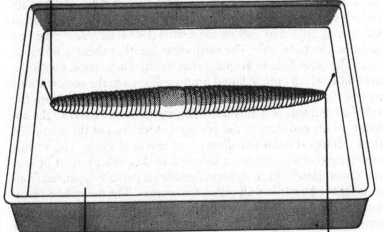

Dissecting pan | Dissecting pin through anal segment

Figure 4. Specimen Arranged for Dissection.

Step 1. With forceps gently lift skin at about 2″ (5 cm) from clitellum.

Step. 2. With razor blade make slight cut at point A.

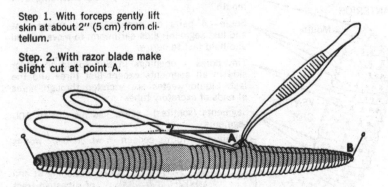

Step. 3. Insert sharp end of scissors. Cut through skin toward anus, slightly to one side of dorsal midline, to point B. Be careful to cut through skin only, no deeper.

Step 4. Make similar cut in opposite direction until you get to about 1″ (2.5 cm) from clitellum.

Figure 5. Exposing Internal Structures of the Earthworm.

12

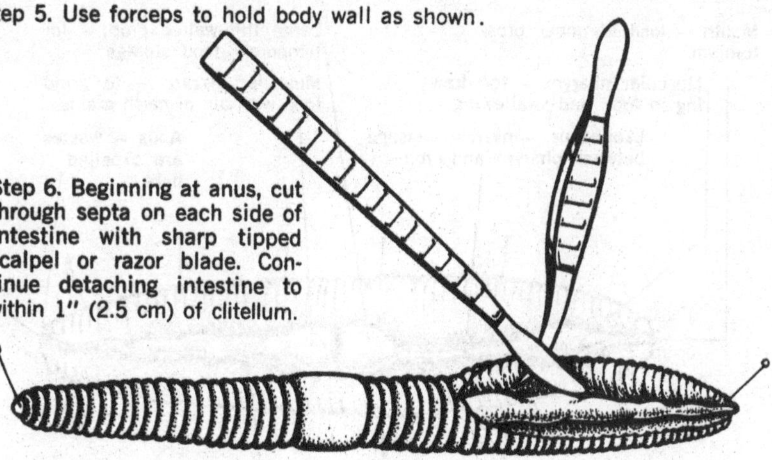

Step 5. Use forceps to hold body wall as shown.

Step 6. Beginning at anus, cut through septa on each side of intestine with sharp tipped scalpel or razor blade. Continue detaching intestine to within 1" (2.5 cm) of clitellum.

Figure 6. Cutting Septa on Each Side of Intestine.

Clitellum

Cut septa   Intestine

Septa cut in posterior segments.
Note arrangement of dissecting pins.

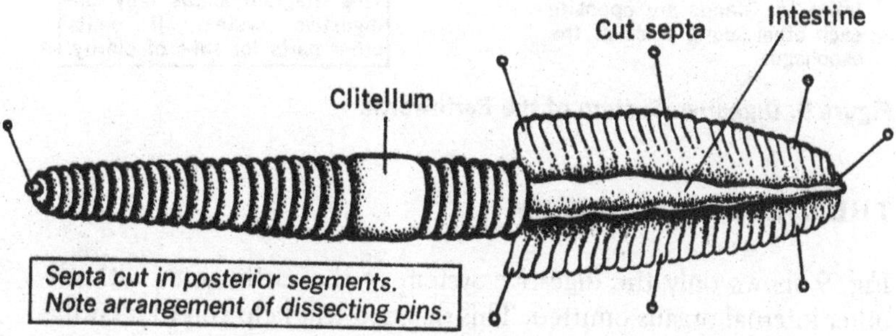

Figure 7. Intestinal Region of the Earthworm—Dorsal View.

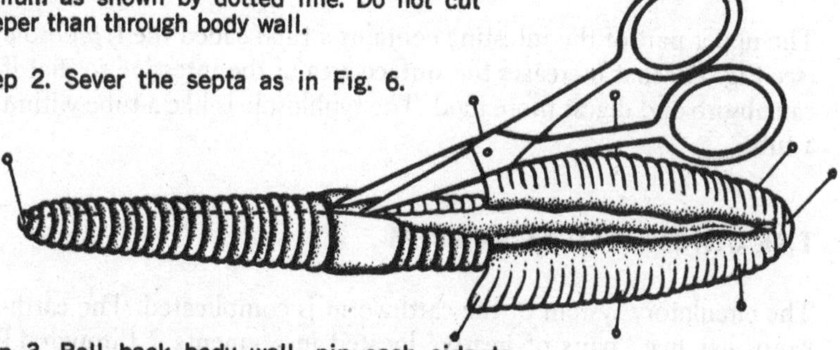

Step 1. Cut through clitellum and up to prostomium as shown by dotted line. Do not cut deeper than through body wall.

Step 2. Sever the septa as in Fig. 6.

Step 3. Roll back body wall; pin each side to dissecting pan to expose all organs as in Fig. 9.

Figure 8. Exposing Internal Organs from Clitellum to Mouth.

13

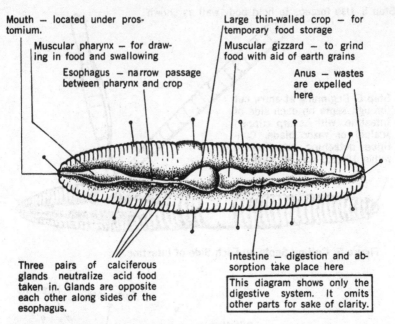

Mouth — located under prostomium.

Muscular pharynx — for drawing in food and swallowing

Esophagus — narrow passage between pharynx and crop

Large thin-walled crop — for temporary food storage

Muscular gizzard — to grind food with aid of earth grains

Anus — wastes are expelled here

Three pairs of calciferous glands neutralize acid food taken in. Glands are opposite each other along sides of the esophagus.

Intestine — digestion and absorption take place here

This diagram shows only the digestive system. It omits other parts for sake of clarity.

Figure 9. Digestive System of the Earthworm.

THE DIGESTIVE SYSTEM

Fig. 9 shows only the digestive system of the earthworm, with all other internal organs omitted. This diagram will help you to identify the parts of the digestive system in the specimen.

TYPHLOSOLE AND INTESTINE

The upper part of the intestine contains a tube called the *typhlosole* (see Fig. 10) that increases the surface area of the intestine so that it can absorb and digest more food. The typhlosole is like a tube within a tube.

THE CIRCULATORY SYSTEM

The circulatory system of the earthworm is complicated. The earthworm has five "pairs of hearts" located in segments 7 through 11 (see Fig. 11). They pump the blood by muscular contractions. The hearts may be seen pulsing in a living earthworm if the worm is held up to the light and gently squeezed. The animal has several large

14

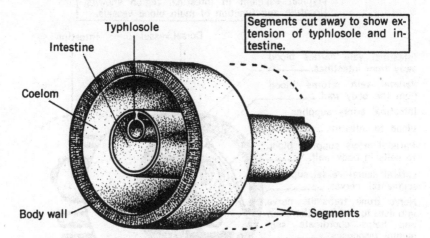

Typhlosole

Intestine

Coelom

Body wall

Segments

Figure 10. Arrangement of Typhlosole Within Intestine—Schematic Diagram.

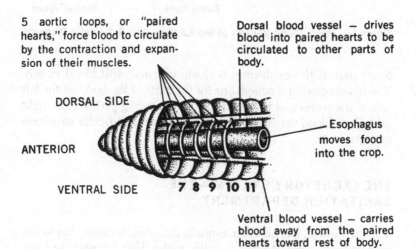

5 aortic loops, or "paired hearts," force blood to circulate by the contraction and expansion of their muscles.

Dorsal blood vessel — drives blood into paired hearts to be circulated to other parts of body.

DORSAL SIDE

ANTERIOR

VENTRAL SIDE   7 8 9 10 11

Esophagus moves food into the crop.

Ventral blood vessel — carries blood away from the paired hearts toward rest of body.

Figure 11. The Paired Hearts of the Earthworm.

blood vessels and many smaller blood vessels. It also contains numerous microscopic blood vessels called *capillaries.*

Study the diagrams in Fig. 11 and in Fig. 12 to locate the hearts and blood vessels in the dissected specimen. The larger blood vessels are readily identifiable; smaller vessels such as the *integumental vessels,* which supply the circulatory needs of the skin, are hard to find.

15

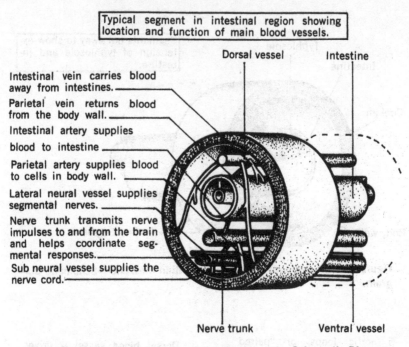

Typical segment in intestinal region showing location and function of main blood vessels.

Dorsal vessel

Intestine

Intestinal vein carries blood away from intestines.

Parietal vein returns blood from the body wall.

Intestinal artery supplies blood to intestine

Parietal artery supplies blood to cells in body wall.

Lateral neural vessel supplies segmental nerves.

Nerve trunk transmits nerve impulses to and from the brain and helps coordinate segmental responses.

Sub neural vessel supplies the nerve cord.

Nerve trunk

Ventral vessel

Figure 12. Major Blood Vessels of the Earthworm—Schematic Diagram.

Every part of the earthworm is richly supplied with blood vessels. The blood vessels are paired, one for the parts of the body on the left side of the worm and the other for corresponding parts on the right side. Each blood vessel carries blood to or from particular structures or organs.

THE EXCRETORY SYSTEM—THE SANITATION DEPARTMENT

The earthworm, like all other animals, including humans, has to dispose of three types of wastes—solid matter, liquid wastes, and gaseous wastes. The solid wastes or feces are expelled through the anus. The liquid wastes and part of the gaseous wastes are removed from the coelomic cavity with the aid of special coiled pipes called nephridia (see Fig. 13). One coiled pipe is a *nephridium*.

Nephridia in the earthworm are found in all but the first three segments and in the last one. Each segment contains a pair of nephridia. You may easily identify them in your specimen by examining the rear, or posterior, segments, where there are few structures to confuse the dissector.

16

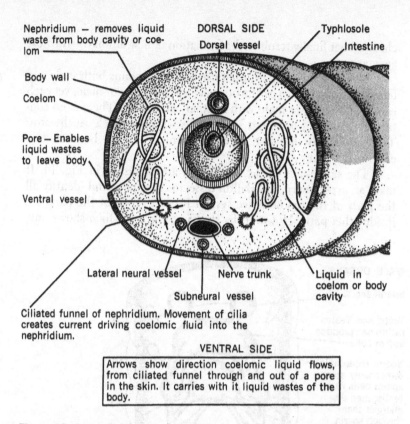

Nephridium — removes liquid waste from body cavity or coelom

DORSAL SIDE
Dorsal vessel

Typhlosole

Intestine

Body wall

Coelom

Pore — Enables liquid wastes to leave body

Ventral vessel

Lateral neural vessel

Nerve trunk

Liquid in coelom or body cavity

Subneural vessel

Ciliated funnel of nephridium. Movement of cilia creates current driving coelomic fluid into the nephridium.

VENTRAL SIDE

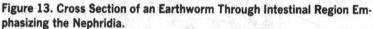

Arrows show direction coelomic liquid flows, from ciliated funnel through and out of a pore in the skin. It carries with it liquid wastes of the body.

Figure 13. Cross Section of an Earthworm Through Intestinal Region Emphasizing the Nephridia.

Examine the diagram of the nephridia in Fig. 13 and trace a nephridium in the specimen from its opening in one segment to its connection with a pore in the skin of the next segment. Use your magnifying lens to explore the nephridial or excretory system. Note that the nephridium has one opening at the ciliated funnel which collects liquid wastes from the coelom and another opening at the pore through which liquid wastes leave the body.

THE REPRODUCTIVE SYSTEM—STARTING THE NEXT GENERATION

The earthworm, like most animals, has sex organs. The main function of these organs is to produce sex cells—sperm cells by the male organs and egg cells by the female organs. The union of a sperm cell with an egg cell starts a number of cell divisions and chemical

17

changes that finally result in the formation of new animals—the next generation.

Unlike many animals, each earthworm contains both male and female sex organs. No earthworms are solely female or male, yet each earthworm needs to mate with another earthworm. There is no contact between egg cells and sperm cells within a single earthworm. The sperm cells are not discharged unless mating between two earthworms takes place.

The sex organs of the earthworm are described in Fig. 14. It takes an experienced and skillful dissector to expose and identify all the parts of the reproductive system because some of the parts are inside other parts. However, by studying Fig. 14, which shows only

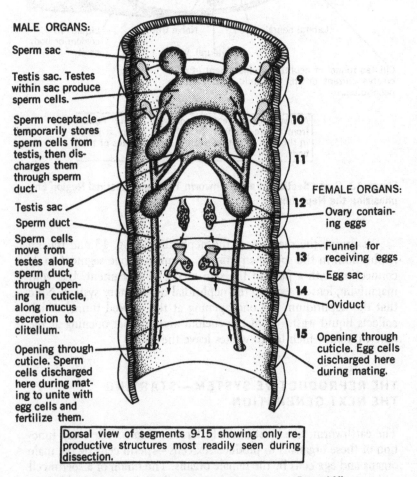

MALE ORGANS:

Sperm sac

Testis sac. Testes within sac produce sperm cells.

Sperm receptacle temporarily stores sperm cells from testis, then discharges them through sperm duct.

Testis sac

Sperm duct

Sperm cells move from testes along sperm duct, through opening in cuticle, along mucus secretion to clitellum.

Opening through cuticle. Sperm cells discharged here during mating to unite with egg cells and fertilize them.

9

10

11

12

13

14

15

FEMALE ORGANS:

Ovary containing eggs

Funnel for receiving eggs

Egg sac

Oviduct

Opening through cuticle. Egg cells discharged here during mating.

Dorsal view of segments 9-15 showing only reproductive structures most readily seen during dissection.

Figure 14. Reproductive System of the Earthworm—Dorsal View.

18

the main parts, you will be able to get a good idea of the structure of the reproductive system.

THE CLITELLUM AS A NEST

The clitellum, which is formed around several segments in preparation for the process of reproduction, forms a cocoon that encases the developing worms. Later in the reproductive process the clitellum is slipped off the body of the worm much as a ring is slipped off a finger. After it has slipped over the anterior end of the worm, a mucus secretion from the skin cells seals each end of the clitellum, forming a protective nest for the developing worms until they are ready to leave their temporary home.

THE NERVOUS SYSTEM—KEEPING IN TOUCH

It may be surprising to learn that the lowly worm has a brain. Of course, it is not as magnificent or as complex as a human brain, but it does enable the earthworm, to a limited degree, to learn from its experiences. Scientific researchers have proved this by testing the behavior of earthworms in a maze, a chamber with pathways designed to lead to an exit (escape or reward chamber) or to one or more dead ends (see Fig. 17 in a few pages).

Just what is a brain? It is a center of nerve cells that helps to coordinate the activities of an animal. In the earthworm the brain is connected with the *double ventral nerve cord* (see Fig. 15). The brain and trunk lines have nerves that branch off to all parts of the body. Thus, any part of the worm affected by something outside itself may send a message to other parts of the body, such as the segmental muscles and the setae. The worm can then retreat to the safety of its burrow. This communication system helps the worm to survive, that is, to get food and to escape its enemies.

In any animal the nervous system is the most difficult to dissect. This is also true for the earthworm. Fig. 15 is a simplified diagram of the worm's nervous system. Study this diagram. Then, with the aid of your magnifying lens or dissecting microscope, locate the labeled parts in your specimen.

At the beginning of the dissection (Fig. 5) you were cautioned not to cut too deeply into the specimen, especially in the anterior region. Now you can see why this is so important. It is very easy to destroy the brain or the nerve centers by careless dissection.

19

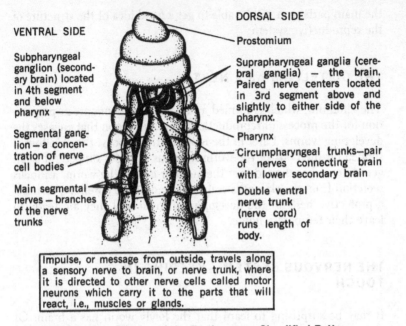

VENTRAL SIDE

Subpharyngeal ganglion (secondary brain) located in 4th segment and below pharynx

Segmental ganglion — a concentration of nerve cell bodies

Main segmental nerves — branches of the nerve trunks

DORSAL SIDE

Prostomium

Suprapharyngeal ganglia (cerebral ganglia) — the brain. Paired nerve centers located in 3rd segment above and slightly to either side of the pharynx.

Pharynx

Circumpharyngeal trunks—pair of nerves connecting brain with lower secondary brain

Double ventral nerve trunk (nerve cord) runs length of body.

Impulse, or message from outside, travels along a sensory nerve to brain, or nerve trunk, where it is directed to other nerve cells called motor neurons which carry it to the parts that will react, i.e., muscles or glands.

Figure 15. Nervous System of the Earthworm—Simplified Pattern.

The organ systems in the earthworm are much simpler than those of animals more commonly used in projects and experiments, such as frogs, mice, or guinea pigs. Therefore the results obtained in experiments with earthworms are often more easily understood than are results from experiments with higher animals.

Earthworms are excellent for many kinds of projects and experiments. They are relatively easy to keep alive in the home or in the laboratory. They take up little space and do not create a mess that requires frequent cleaning. As long as they are kept moist there is little danger of injuring the worms by handling. Here are some interesting experiments and projects to try with earthworms.

Project 1: Demonstrating the Beating Hearts of the Earthworm

Obtain two 6-inch (15-cm) squares of ordinary windowpane glass. Place a ¼-inch (6-millimeter, or 6-mm) layer of vaseline along all four edges of one glass square. Place a live earthworm within the vaseline-edged square as shown in Fig. 16, and follow the instructions in Fig. 16. When you have completed this experiment, examine the specimen between the glass squares with your hand lens or dissecting microscope to get an enlarged view of the beating hearts.

Many different and exciting experiments may be developed using the same technique as that used to demonstrate the beating hearts. For example, you can study the effects of hormones such as adrenin (adrenalin) and insulin on the heartbeat of the worm. Keep the worm for a short time on a moist surface, such as a blotter, saturated with a solution of the substance whose effects are being studied. Then transfer the worm to the glass squares and observe the effects of the chemical on the worm. In this way you can study the possible effects of such chemical substances as aspirin, alcohol, ascorbic acid, glutathione, thyroid extract, auxin, gibberellic acid, etc., on the circulatory system of the earthworm.

Project 2: Investigating the Sex Cells of an Earthworm

The sperm cells of the earthworm can be studied under a microscope having at least 400× magnification. This project requires slides, cover slips, a *Syracuse dish* or *Petri dish* and a medicine dropper, as well as dissecting instruments.

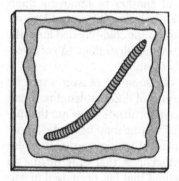

Step 1. Prepare glass square with vaseline along edges as shown.

Step 2. Place live earthworm within vaseline edges of square. Vaseline forms a seal between 2 glass squares (see Step 3) and prevents worm from escaping.

Step 3. Quickly cover earthworm with second glass square to hold worm between the 2 squares.

Step 4. Gently press glass squares together to squeeze worm's body into an almost oval shape.

Step 5. Hold squares in front of strong light to see beating hearts of the worm.

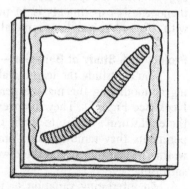

Figure 16. Method of Studying Beating Hearts of the Earthworm.

Obtain live earthworms from a biological supply house or collect them yourself. Live worms are most easily collected after a heavy rain when they come out of their burrows. The worms can be put under anaesthesia by immersing them in a 5% urethane–water solution for 5 minutes. Dissect out the sperm sacs and testes by following the procedure described in Figs. 5, 6, 7, and 8. (Consult Fig. 14 to help locate the sperm sacs and testes.) Place them in the Syracuse dish or Petri dish. Add a few drops of water to the dish. Break up the structure of the sperm sacs and testes with a pair of dissecting needles. Now draw up several drops of the mixture with a medicine dropper and transfer one drop to a microscope slide. Examine the drop with the microscope to find the living sperm cells. This whole process should be accomplished quickly because the cells do not remain alive for long.

You can also examine the egg cells of the earthworm under a microscope. Follow the same procedure as described for obtaining sperm cells but dissect out the ovaries instead. Consult the diagram of the reproductive system (Fig. 14) to help you locate the ovaries.

It would be most interesting to try to fertilize the egg cells of one earthworm with the sperm cells of another by removing them from earthworms as described above and placing the egg and sperm cells in a dish containing Ringer's solution for cold-blooded animals. Consult your science teacher or any text on physiology to get information on the preparation of Ringer's solution.

Studies of the reproduction and development of worms may be done by dissecting the clitellums of worms of different lengths. Small worms are generally not as likely to have fertilizations as are the mature worms. A study of the cocoons of worms may be made, to observe different stages in the development of young worms. It is also possible to do experiments on *parthenogenesis*, that is, fertilizing an egg cell *without* a sperm cell. One method that has been used in parthenogenesis experiments is to pierce the egg's membrane gently with a fine glass needle or a metal needle.

Project 3: A Study of Behavior—The Worm Turns
You can study the learning ability of the earthworm by repeating or modifying the maze experiments performed by Yerkes and Heck (see Fig. 17). They discovered some interesting things about the earthworm's ability to profit from experience. In a series of experiments they removed the "brain" or cerebral ganglia of earthworms. When they tested these worms in the maze the results were surprising!

An interesting variation of the maze experiment would be a study of the reactions of the earthworm in a similar maze, using a

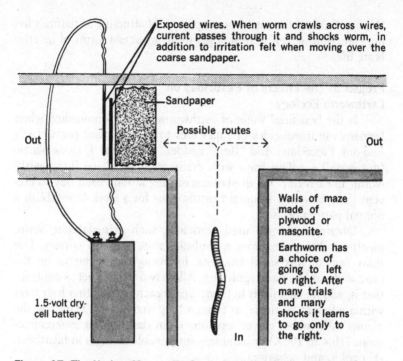

Exposed wires. When worm crawls across wires, current passes through it and shocks worm, in addition to irritation felt when moving over the coarse sandpaper.

Sandpaper

Possible routes

Out

Out

Walls of maze made of plywood or masonite.

Earthworm has a choice of going to left or right. After many trials and many shocks it learns to go only to the right.

1.5-volt dry-cell battery

In

Figure 17. The Yerkes Maze—To Study Behavior of the Earthworm.

blotter saturated with a harmless acid like acetic acid or vinegar instead of the electric shock method. No doubt you will think of many other interesting behavior experiments with earthworms.

Project 4: We Can Cultivate Earthworms

It is quite easy to maintain earthworms in a special bedding material called "Buss" bedder—a commercial product. Once you have succeeded in raising some worms, it is possible to introduce various chemicals into the bedding compound to test the effects of these chemicals on the growth, reproduction, and behavior of the earthworm. Test the effects of such chemicals as tryptophan, nucleic acid, thyroxin, cytosine, riboflavin, alpha tocopherol (vitamin E), ascorbic acid, and many others.

The earthworm is made up of many similar segments with little specialization to differentiate one segment from another. It does not even have a distinct head. In higher animals you will see the segments become specialized, with well-developed appendages and definite body divisions, including a head equipped with a complicated brain and sense organs. Thus the earthworm presents the simplified evolutionary pattern of the structure of higher invertebrate animals.

We can think of the earthworm's pattern of structure as nature's living clay, from which were molded more advanced forms of invertebrate life.

Project 5: The Effects of Pesticides on Earthworm Ecology

Is the beneficial value of earthworms lost or diminished when humans use insecticides or fungicides to control plant pests? Let's find out. Procedure: Add "Buss" bedder (see Project 4) to six plastic (not metal) seedling trays with drainage holes. Place three earthworms in each tray. Wrap plastic screening around each tray to prevent their escaping. Maintain earthworms for a week to establish a normal pattern of life.

Obtain commonly used pesticides, such as malathion, sevin, nicotine sulfate, etc., from a hardware store or plant nursery. Use spray cans or mix small amounts, following the directions on the cans, and put in a hand spray gun. Allow two trays to act as controls, that is, add no chemicals to them. Spray each of the other four trays with a different pesticide, as though they were plants. Handling the worms with rubber gloves, examine them daily with a stereomicroscope (10× to 100×) or hand lens for possible changes in heartbeat, skin color, and behavior.

Select one worm from each tray every third day, immerse in alcohol, and perform an autopsy. Look for comparative size of organs such as gonads and intestine. Evaluate results.

3

Animals with Armor

THE CRAYFISH

IN CHAPTER 2 WE DISSECTED A SIMPLE, SEGMENTED ANIMAL, THE earthworm, a creature with no protective armor. Now we shall work with a crayfish, an animal related to the simple earthworm. The crayfish, however, has specialized segments with a tough suit of armor and external appendages, or limbs. It uses its limbs for fighting, for getting food, for locomotion, and so on.

A crayfish enlarged to the size of a dinosaur like the *Tyrannosaurus* would appear far more terrifying than the dinosaur. Imagine a creature with gigantic crushing pincers, with long waving antennae searching for victims, with powerful crushing jaws—a monster equipped with a thick wall of heavy armor in the form of an outer skeleton. Fortunately the crayfish grows to only about 5 inches (12 cm) in length. Its relative, the lobster, can grow about 2 feet (61 cm) long and be quite formidable. Fig. 18 shows a crayfish.

The crayfish is a scavenger. It lives on the muddy bottoms of streams and ponds, emerging at night to feed on dead matter, live

Figure 18. Large Crayfish *(Cambarus)*. (Photograph by Carolina Biological Supply Company)

25

insect larvae, and worms. It is often cannibalistic. Raccoons, freshwater bass, muskrats, crows, and humans are among its mortal enemies. To escape its foes it flexes its tail, the *uropod*, spreads it like a fan, and draws it forward underneath its body with a strong motion. This causes the crayfish to dart backward into the muddy bottom, which is further stirred by its rapid motion. Its enemies generally lose it in the muddied water.

The tiny crayfish can cause enormous damage. It burrows several feet deep into the bottoms of streams, especially near overhanging banks. These innumerable burrows, frequently enlarged by muskrats, weaken dikes or levees, and serious floods may develop.

Some gill-breathing cousins of the crayfish are lobsters, crabs, shrimps, prawns, barnacles, and krill. The *Crustacea*, the group to which all of these belong, are an important food source for humans. Lobsters, crabs, shrimps, prawns, and crayfish are eaten by humans. Prawns resemble shrimps except for a little hump on their *cephalothorax* or back. In fact the "shrimp" that you eat in a shrimp salad may have been sold to your local fish market by a "prawn broker." You would not know the difference.

Barnacles, a variety of crustacea, do great damage to our ships and wharves. A shrimplike crustacean commonly called krill or whale food is estimated to produce over a billion tons of food for whales each year. Scientists have been thinking of using krill to supplement the food resources of man.

The outer skeleton or *exoskeleton* of the crayfish and of other invertebrates consists mainly of a tough substance called *chitin* (see Project 1 in this chapter). It is a very effective protection, but it also imprisons the growing crayfish. In order to grow, the crayfish must escape from its exoskeletal prison. This it does by molting, or shedding, its outer skeleton. The exoskeleton splits at a crease in the cephalothorax (see Fig. 19) and the crayfish squeezes out of the exoskeleton, sometimes leaving behind a leg as well as the shell. Thus defenseless, it hides until it grows a larger exoskeleton. Often, to escape an enemy, the crayfish will snap off its own leg if the enemy is holding it. The crayfish has the power to regenerate, or grow back, another leg if it has lost one.

The crayfish can walk backward, forward, and even sideways because it has seven joints in each walking leg, each joint arranged in a different position. Abdominal appendages called *swimmerets* help to keep the animal clean. The first two swimmerets in the male are modified to form channels for transporting sperm cells to the female. In the female the eggs, which look like clusters of grapes, are attached to the swimmerets at egg-laying time. They remain attached to the outside of the swimmerets until the developing crayfish have

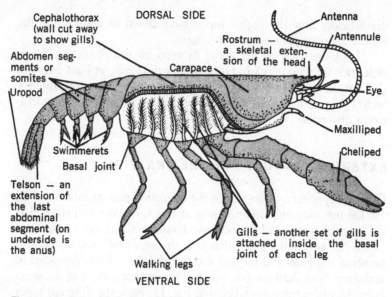

Cephalothorax
(wall cut away
to show gills)

Abdomen seg-
ments or
somites

Uropod

DORSAL SIDE

Rostrum —
a skeletal exten-
sion of the head

Carapace

Antenna

Antennule

Eye

Maxilliped

Cheliped

Swimmerets

Basal joint

Telson — an
extension of
the last
abdominal
segment (on
underside is
the anus)

Walking legs

Gills — another set of gills is
attached inside the basal
joint of each leg

VENTRAL SIDE

Figure 19. External Anatomy of the Crayfish.

grown large enough to achieve independence from the mother cray-fish.

Like the earthworm, the crayfish is built on the segment plan. However, the crayfish is a more advanced animal than the earthworm. Therefore, its segments are not all alike. While the earthworm has primitive structures called setae, the crayfish has specialized appendages, such as swimmerets, walking legs, antennae, etc. (see Fig. 19). Instead of a pair of nephridia in each segment, as in the earthworm, we find in the crayfish a single excretory organ. The nervous system and the digestive system in the crayfish resemble those of the earthworm. But unlike those of the earthworm, the sense organs of the crayfish, such as the eyes and antennae, are well developed. The earthworm has only light-sensitive spots in its head and no antennae, while the crayfish has large compound eyes, each eye made up of about 2,500 sections, and large antennae resembling those of the grasshopper.

PREPARING TO DISSECT THE CRAYFISH

The crayfish dissection will show how the structure of a more complicated invertebrate, like the crayfish, is built on the simple foundation of the earthworm's plan of structure. It will also show dramatically how the higher forms have added to and altered the simple segment plan of the earthworm.

Let's see what we are going to need for this dissection. First, order two injected crayfish, type *Astacus fluviatilis*, at least 3 inches (7.5 cm) long. Then obtain a pair of sharp, curved scissors about 4 inches (10 cm) long, and the same equipment as listed for the earthworm dissection.

EXTERNAL FEATURES OF THE CRAYFISH

Place the specimen on its side in the dissecting pan, as shown in Fig. 19. Do not use any dissecting pins at this time. We will study first the external features of the crayfish. Examine the dorsal surface and locate the two main body divisions, the *cephalothorax* (fused segments of the head and thorax) and the *abdomen*. Note segments in abdomen. Now examine one side and the ventral surface of the specimen. Identify structures labeled in Fig. 19. Note the different types of appendages—antennae, antennules, mouth parts, legs, and swimmerets.

EXPOSING THE GILLS

Before exposing the crayfish's gills, remove the legs and swimmerets from the abdomen and thorax regions by cutting through the first or basal joint near the body. Follow directions in Fig. 20 and cut away a section of the *carapace* (exoskeleton covering the head–thorax region).

How does the attachment of the gills to the legs help the crayfish in breathing? (See Fig. 21.) When the legs move, the gills attached to the legs move in a waving motion like a flag on a flagpole. This waving motion stirs the water which enters under the carapace. Since water contains dissolved oxygen (O_2), the stirring of the water brings the oxygen to the blood canals in the gills where absorption of oxygen into the bloodstream takes place.

DORSAL VIEW OF THE CRAYFISH

You have just examined the crayfish from a side view. Now let us study its dorsal side. We can expose several important organs and blood vessels by cutting a long, narrow section of the exoskeleton

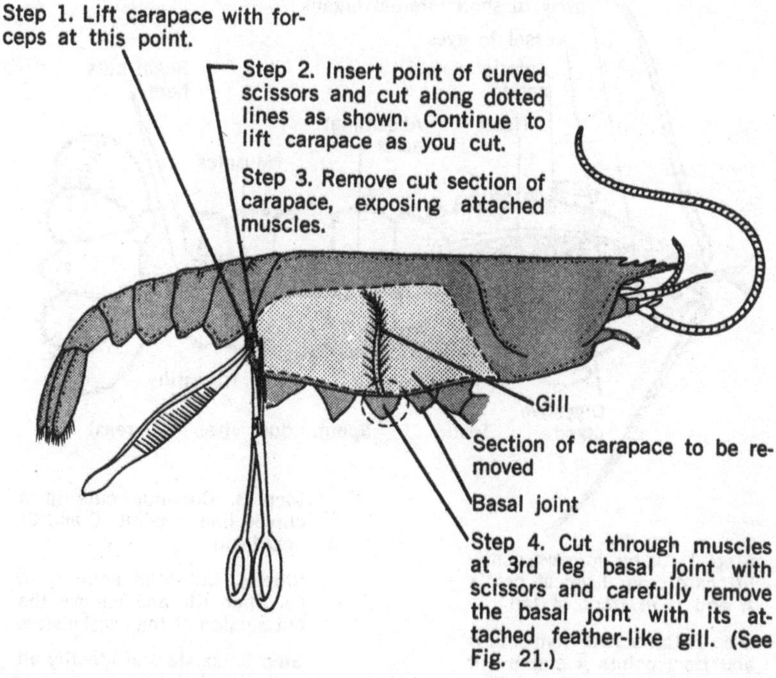

Step 1. Lift carapace with forceps at this point.

Step 2. Insert point of curved scissors and cut along dotted lines as shown. Continue to lift carapace as you cut.

Step 3. Remove cut section of carapace, exposing attached muscles.

Gill

Section of carapace to be removed

Basal joint

Step 4. Cut through muscles at 3rd leg basal joint with scissors and carefully remove the basal joint with its attached feather-like gill. (See Fig. 21.)

Figure 20. Exposing Gills of the Crayfish.

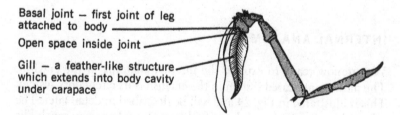

Basal joint — first joint of leg attached to body

Open space inside joint

Gill — a feather-like structure which extends into body cavity under carapace

Figure 21. Walking Leg of the Crayfish Showing Gill Attachment.

from the dorsal side of the crayfish. Place the crayfish in the dissecting pan, dorsal side up. Now study Fig. 22 and follow directions for the dissection.

Do not remove the mouth parts until you have studied the organ systems. (See Figs. 22, 23, and 24 for locating these systems— the nervous, excretory, and circulatory systems.) You may damage the organ systems unless you acquaint yourself with their relation to the mouth parts.

29

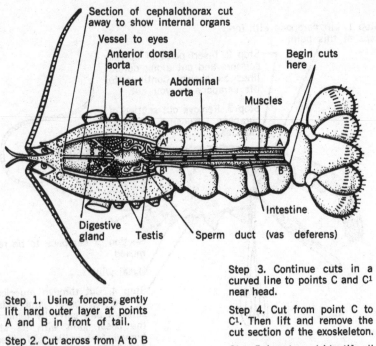

Step 1. Using forceps, gently lift hard outer layer at points A and B in front of tail.

Step 2. Cut across from A to B and from points A and B cut forward to points A¹ and B¹.

Step 3. Continue cuts in a curved line to points C and C¹ near head.

Step 4. Cut from point C to C¹. Then lift and remove the cut section of the exoskeleton.

Step 5. Locate and identify all parts labeled here.

Figure 22. Dorsal View of Internal Organs of the Crayfish.

INTERNAL ANATOMY

We are now ready to explore the internal anatomy of the crayfish. The most complicated system in the crayfish is its circulatory system. This is illustrated in Fig. 24 and will be described in detail later. The other systems of the crayfish are fairly simple and are very much like the systems in the earthworm, except for the excretory system. All the main systems, except the circulatory system, are shown in Fig. 23.

EXPOSING THE MAIN ORGAN SYSTEMS

Using small curved scissors, cut away the basal joints to which the legs were attached (see Fig. 20). Now remove with scissors all gills in the section of the cephalothorax which was previously opened. Examine the gills with a hand lens to observe how they are adapted for

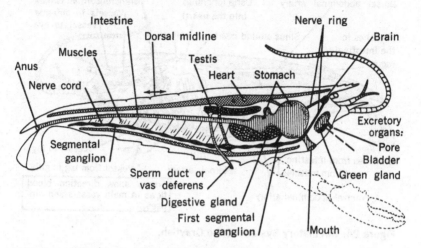

Figure 23. Side View of Internal Organs of the Crayfish.

getting oxygen into the blood. They are very thin and feathery, offering a great deal of surface for taking in oxygen from the water.

Carefully lift and cut away hard outer protective plates from dorsal midline to base of each leg. Use forceps to lift and hold the plates from the dorsal midline as you cut to the base of each leg. Use forceps and probe to remove masses of muscle from the region of the eyes down to the last walking leg on the side being dissected. Then cut away the skeletal plates down to the tail to expose the abdomen. As you lift and cut muscles, watch to see whether you are cutting into any other organs. Avoid damaging any parts of the systems shown in Fig. 23.

With a fine probe, trace the body systems in the following order: (1) the excretory system (near the mouth), (2) the digestive system, (3) the reproductive system, and (4) the nervous system.

How does the nervous system of the crayfish compare with the nervous system of the earthworm?

THE CIRCULATORY SYSTEM

Now that the crayfish has been opened we can study the circulatory system and locate the heart and the main blood vessels. Arrows in Fig. 24 show direction of blood flow to and from the heart. Locate in the specimen the main vessels labeled in Fig. 24. There are no veins in the crayfish. Blood flows from heart to arteries to capillaries and then into tissue spaces called *sinuses* which serve in place of veins.

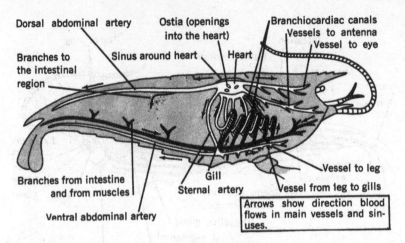

Figure 24. Circulatory System of the Crayfish.

The sinuses around the gills are called canals. Blood sinuses are also found in insects. Man has blood sinuses in the liver. He has hollows in the facial bones which are also called sinuses. If these become infected, we suffer from what is popularly known as a "sinus attack."

APPENDAGES

The word "appendages" generally means limbs, such as arms, wings, legs, or fins. Scientifically, appendages are all movable structures that extend outward from the body of an animal regardless of the part of the animal to which the extension is attached. In the crayfish the antennae, antennules, and other parts attached to the head and abdomen are appendages. Each has an important function, and all appendages are controlled by the nervous system. They govern the senses of touch and taste and they help the crayfish to breathe and to chew its food.

We shall now dissect and examine the appendages of the head (see Fig. 25). First free the mouth parts from the head with sharp-tipped scalpel, scissors, and forceps. Then dissect out the antennae by holding them with forceps and cutting a shallow groove into the head where the antennae are attached. This will free the antennae. Find the nerve and blood vessel attachments to these parts. The nerves look like thin, white cords. Study the mouth parts with a hand lens or dissecting microscope, and with a probe find out how they are arranged to carry out their functions.

Use a hand lens or dissecting microscope to study appendages.

Study the table in Fig. 25 to understand how the design of each structure helps it to do its job effectively.

With a hand lens find the blood vessels in a detached walking leg. Note how the gill is attached to the leg (see Fig. 21). The open space in the leg is part of a blood sinus.

Now that you've come this far, try your hand at the following projects. The purpose of these projects is to provide interpretive experiences for the reader, experiences that can only be really fulfilled by actual comparative dissections and by the analysis of comparable parts.

Project 1: Behind the Chitin Curtain

Chitin is the horny substance that forms the hard outer shell of the crayfish and of other crustaceans and insects. Let's find out what's behind it. We can begin to understand the evolutionary changes in the structure of crustaceans by making a comparative study of the lobster, shrimp, and barnacle.

Obtain injected specimens of these animals. Compare their external organization with that of the crayfish. Do the same with the internal organization, after the specimens have been dissected. If you run into difficulty, you might consult your biology teacher or refer to a book on invertebrate zoology.

Assemble the mouth parts of each specimen and show how they compare in structure and function with the mouth parts of the crayfish. Observe how the crustacean pattern of structure has been modified in the barnacle. How have these changes helped the barnacle to survive?

Project 2: Making a Photographic Study of the Appendages

Many science magazines are eager to get unusual nature pictures for their natural history features. They pay well. Perhaps you can do some profitable freelance nature photography.

Dissect out and photograph the appendages of crustaceans (limbs, mouth parts, etc.). Arrange these photographs in order so that similar structures in the different animals such as mandibles, appendages, and so on, may be compared. For instance, the photographs of the mandibles of these organisms should be set side by side.

Project 3: New Parts for Old

Do you want to create new animals from old parts? Try your hand at experiments on regeneration. Regeneration is the ability of some animals to grow back a lost part of the body. Animals which have been used successfully in experiments on regeneration are tadpoles, salamanders, earthworms, planaria, crayfish, and hydra.

APPENDAGES OF CEPHALOTHORAX (HEAD AND THORAX)

	APPENDAGE	FUNCTION	LOCATION
	Antennules.	Detects touch and taste. Helps to balance crayfish.	Front of mouth.
	Antenna.	Detects touch and taste.	Front of mouth.
	Mandible or jaw.	Crushes food.	Mouth.
	First maxilla.	Moves food to mouth.	Behind mandibles.
	Second maxilla.	Bails water in gill chamber.	Behind mandibles.
	First maxilliped.	Holds food. Touch. Taste.	At forward and ventral part of thorax region.

	Name	Function	Location
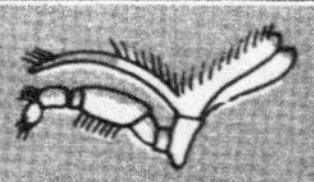	Second maxilliped.	Holds food. Touch. Taste.	At forward and ventral part of thorax region.
	Third maxilliped.	Holds food. Touch. Taste.	At forward and ventral part of thorax region.
	Walking leg, four on each side.	For locomotion.	Posterior to maxillipeds at ventral part of thorax.
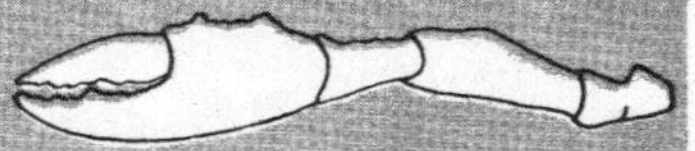	Cheliped, the first leg.	To grasp food.	Posterior to maxillipeds at ventral part of thorax.

APPENDAGES OF THE ABDOMEN

	Name	Function	Location
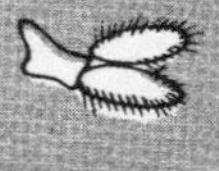	Swimmeret.	First swimmeret in male transfers sperm to female who uses 2nd, 3rd, 4th and 5th swimmerets to hold eggs and young.	Abdominal region on ventral side.
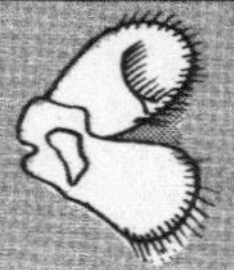	Uropod.	For swimming.	Tail end.

Figure 25. Appendages of the Crayfish and Their Functions.

Try testing the effects of such chemicals as glutamic acid and acetylcholine on regeneration by adding these chemicals to the water in which the animals are kept. Several biological supply houses have printed material available on the subject of regeneration.

Project 4: Home Life of the Crayfish

Study the crayfish in its natural surroundings or habitat. Record your observations of its movements, the food it eats, and how it reproduces itself. City folks rarely get to see crayfish in their natural surroundings. However, even city folks can study the life of a crayfish in an aquarium.

It is possible to set up an aquarium at home or school and observe the development of young crayfish in it. For your aquarium use pond water, water plants, and a bottom consisting of gravel, earth, and small stones. Arrange the bottom so that there are miniature steep banks. Add maggots and other food to the water. Stock the tank with snails, small freshwater clams, and small fish.

Before we move on to the next dissection, let us stop to consider what we have accomplished. We have learned how to dissect, what instruments to use, and how to use them. We have dissected a simple, spineless animal, the earthworm. This gives us some idea of what a primitive animal is like. We moved onward to dissect the crayfish, an animal linked by distant kinship to the earthworm and by closer kinship to the most advanced forms of invertebrates, the insects. Let's now go on to examine the grasshopper, one example of an advanced form of invertebrate.

4

DANGER: GRASSHOPPER

GRASSHOPPERS ARE FOUND ALMOST EVERYWHERE. THEY WILL EAT practically any wild or cultivated plant. In some areas of the United States special contraptions called hopperdozers have been used to catch grasshoppers in cultivated fields. Hopperdozers have caught as many as a million and a half grasshoppers per acre (1 acre = about 4050 square meters). Just imagine how many grasshoppers there must be in the whole United States!

A grasshopper cannot eat much by itself. But it has been estimated that seventeen grasshoppers per square yard (1 square yard = about 0.84 square meter), on a 40-acre field (160,000 square meters or 16 hectares), can eat 1 ton of alfalfa hay in one day! Multiply that figure by the millions of acres of farm land, and the possible destruction of crops becomes alarming. That is why we started by saying "Danger: Grasshopper."

THE TOP OF THE INVERTEBRATE LADDER—THE INSECTS

The grasshopper belongs to the highest and most complicated group of invertebrate animals, the insects. If we compare our old friend the earthworm with the grasshopper, we can quickly see how much they resemble each other (see Fig. 26). For example, the grasshopper's body, like the body of the earthworm, is arranged in segments. The earthworm has paired appendages (the setae) attached to its segments. The grasshopper also has paired appendages attached to its segments. However, the grasshopper's appendages are naturally more complicated and more specialized.

Figure 26. Wingless Grasshopper *(Romalea microptera).* (Photograph by Carolina Biological Supply Company)

The earthworm appeared on earth much earlier than the grass-hopper. In fact, the earthworm and its kin were at one time the highest form of animal life on earth. In the previous chapter we examined and dissected a crayfish. We could see how the crayfish is descended from the earthworm. Actually, in evolutionary development, the crayfish and its cousins are sandwiched between the primitive earthworm and the advanced insects. The anatomy of the grasshopper shows this evolutionary development more clearly than do many other animals.

Thus far, we have mentioned only the external features that relate the earthworm to the crayfish and the grasshopper. To better understand the close ties among these animals we must penetrate their chitin overcoats (exoskeletons) by dissecting them. We can then see how the different systems, such as the digestive, circulatory, nervous, and other systems of these groups compare with each other.

COUNTDOWN FOR DISSECTION

This dissection requires several injected specimens of the Carolina grasshopper, *Dissosteira carolina,* or *Romalea microptera,* which is a larger type. Get at least two males and two females because this is a difficult dissection. In addition to the equipment used for dissecting the earthworm, we will need a pair of small scissors with fine, straight

38

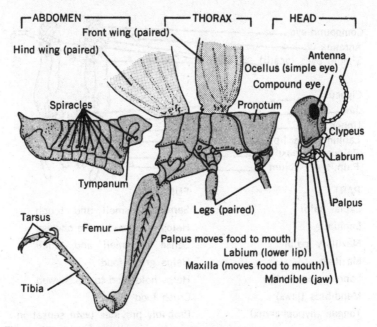

```
┌─ABDOMEN ────────┐        ┌─ THORAX ─┐    ┌─ HEAD ──────┐
```

Front wing (paired)

Hind wing (paired)

Antenna

Ocellus (simple eye)

Compound eye

Spiracles

Pronotum

Clypeus

Labrum

Tympanum

Palpus

Tarsus

Legs (paired)

Femur

Palpus moves food to mouth

Labium (lower lip)

Maxilla (moves food to mouth)

Mandible (jaw)

Tibia

Figure 27. External Anatomy of the Male Grasshopper.

blades about 2½ inches (6.4 cm) long, and a narrow-tipped medicine dropper about 3 inches (7.6 cm) long.

EXTERNAL ANATOMY

Before we dissect the grasshopper, let's study Fig. 27 and acquaint ourselves with the external characteristics. Locate the three main body divisions on your specimen—abdomen, thorax, head. Notice the segmented character of the body. Identify all parts of the abdomen and thorax shown in Fig. 27. Use hand lens to see the spiracles in the abdomen.

How many pairs of legs does the grasshopper have? How many did the crayfish have? A good principle to remember is that the more advanced an animal is, the fewer appendages it has, either in its adult or its embryonic stage.

Examine the wings with a hand lens. As in most insects there are two pair of wings. The arrangement of the veins in the wings is *inherited* and is different in each insect. Examine the head with a hand lens. The head is really a fusion of the first six segments. Identify each part of the head as labeled in Fig. 28.

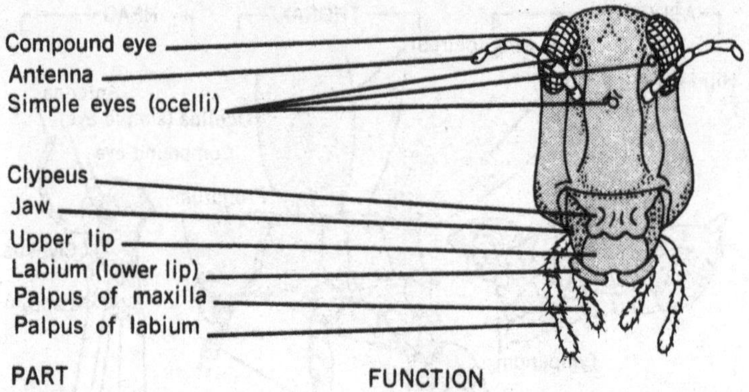

Compound eye
Antenna
Simple eyes (ocelli)

Clypeus
Jaw
Upper lip
Labium (lower lip)
Palpus of maxilla
Palpus of labium

PART	FUNCTION
Labial palpi	Sense of smell and touch
Labium	Helps hold food to be chewed
Maxillary palpi	Sense of smell and touch
Maxillae	Helps grind food
Labrum (upper lip)	Helps hold food to be chewed
Mandibles (jaws)	Crush food
Tongue (hypopharynx)	Probably provides taste sensation

Figure 28. Face View of the Male Grasshopper's Head.

DISSECTING THE GRASSHOPPER

Figs. 29 and 30 give directions for all the steps in dissecting the exoskeleton. The first steps in this dissection are the removal of the legs and wings. This will make it more convenient to handle the body of the specimen as we proceed further. Identify with hand lens the *femur, tibia,* and *tarsus* after you remove the legs. Put wings aside for future study. Let's continue by dissecting the male and the female grasshopper as shown in Fig. 29.

Before removing the mouth parts, trace each part back to its point of attachment (see Fig. 28). Then lift the part with forceps and, using sharp scissors or scalpel, cut each part away at the point of attachment. Begin by removing each *palpus.* Then remove upper lip and jaws (*mandibles*). Finally, remove the tongue (*hypopharynx*), which is beneath the upper lip. To see points of attachment clearly, use your hand lens.

On a piece of paper, arrange the mouth parts that you removed so that they are in the same position they were in, originally, in the grasshopper's head. This will enable you to see how the grasshopper grasps its food and moves it toward its jaws. Note that the grasshopper chews sideways or laterally. How does this compare with most animals?

40

REMOVING THE ANTENNA, COMPOUND EYE, AND PRONOTUM

To remove each antenna, cut with scissors at point of attachment to head. Note joints in the antenna. This is what enables it to move and explore the environment. The antennae have nerve endings sensitive to touch and smell. How does this help the grasshopper to explore its environment? Review Fig. 25 and note differences in appendages of crayfish and of grasshopper.

Use a thin, sharp scalpel to dissect out one compound eye. Cut inwardly along the outline of the eye until the eye is free, and remove it. Note blood vessel and nerve connections extending from the eye. Examine the three *ocelli* or simple eyes with a hand lens. Compare these with the compound eyes. What do you see?

To remove the *pronotum* (outer shield), use forceps and scalpel as shown in Fig. 29.

To remove the exoskeleton of abdomen and head, follow the directions in Figs. 29 and 30.

INTERNAL ANATOMY

It was necessary to remove the appendages and part of the exoskeleton before attempting to examine the internal organs. Study Fig. 31 to identify the main internal organs.

The circulatory system (Fig. 31) is very simple. There is one main vessel that runs along the dorsal midline of the body. Examine the underside of the removed exoskeleton to find the dorsal blood vessel, since it will probably remain attached to the upper part of the abdominal skeleton. The front end of the vessel going to the head is called the dorsal *aorta*. The back part of the vessel has several swellings called hearts. Each swelling (heart) has a tiny opening (*ostium*) equipped with valves that allow blood to enter the heart. When the heart contracts, the valves close and the blood is driven through the vessel toward the head. There the blood passes into a body cavity (*haemocoel*) which is continuous throughout the whole animal. The blood carries digested food, which the surrounding cells absorb. Unlike most animals, however, the blood of the grasshopper has little to do with carrying oxygen. It is interesting to note that the grasshopper has no red blood corpuscles. It does have white blood corpuscles. The blood returns from the haemocoel to the *ostia* (plural of ostium) of the hearts and follows the circulatory route all over again. This type of circulatory system is called an *open system* because the blood flows freely through open tissue spaces. Unlike the grasshopper, man

41

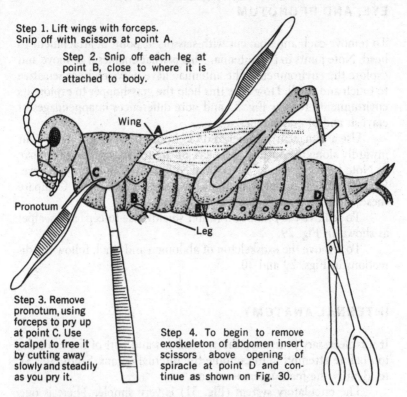

Step 1. Lift wings with forceps. Snip off with scissors at point A.

Step 2. Snip off each leg at point B, close to where it is attached to body.

Wing

A

Pronotum

C

B

Leg

Step 3. Remove pronotum, using forceps to pry up at point C. Use scalpel to free it by cutting away slowly and steadily as you pry it.

Step 4. To begin to remove exoskeleton of abdomen insert scissors above opening of spiracle at point D and continue as shown on Fig. 30.

D

Figure 29. Beginning Dissection of the Female Grasshopper.

has a closed circulatory system in which the blood is always contained in blood vessels.

Find the dorsal aorta and the hearts in Fig. 31. Then examine the upper part of the exposed specimen with a hand lens and locate the dorsal aorta and the hearts. If you do not see the hearts on the specimen, look at the exoskeleton you have cut away. Use a hand lens to see the ostium or pore in each heart.

THE RESPIRATORY SYSTEM

The grasshopper has no gills and no lungs. How then does it breathe? Look first at Fig. 27 and find the spiracles. Now find them on your dissected specimen. You will probably have to dissect another specimen to study the respiratory system because it is usually

42

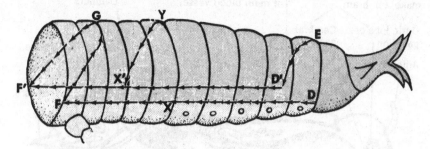

Step 5. Cut from point D to E and continue down on other side to D^1. Make all incisions shallow.

Step 6. Then cut from point D to F and from F to G, and continue down on other side to F^1.

Step 7. Cut from point D^1 to F^1 on the other side of body.

Step 8. Now cut from X to Y and continue down on other side to X^1. The exoskeleton is now divided into 2 cut sections, F to X, and X to D.

Step 9. Carefully lift the sections off with forceps. Use probe and scalpel to separate the sections from adhering muscle.

Step 10. Use forceps and scissors to remove all muscles surrounding the organs. Muscles look like small brown strands of twine. Use hand lens and pen flashlight to prevent damage to organs as you probe and cut.

Step 11. Move reproductive gland aside with probe to see stomach, gastric pouches and excretory tubules.

Step 12. To see ganglia and double nerve trunk, raise entire digestive system by sliding probe under it and pressing gently upward. Nerve trunk is on midline resting on ventral exoskeleton. Pick away muscles with forceps to expose nerve trunk.

Step 13. With scissors remove exoskeleton and underlying muscles from top of head between eyes and up to antennae. (Do not cut off the head.) This will expose the "brain."

Figure 30. Removing Exoskeleton to Expose Internal Organs.

damaged during dissection unless great care is exercised. Place the narrow tip of the medicine dropper or the needle tip of a hypodermic syringe firmly against the opening of a spiracle. Squeeze the rubber ball gently. Hold a hand lens over the side of the abdomen and observe the slight strain or swelling on a saclike structure called the abdominal air sac (see Fig. 32). This may not work with the first spiracle because the spiracular valve may be closed. If so, try another spiracle. When the grasshopper breathes in, the valves in the first four pairs of spiracles are open and the valves in the last six pairs of spiracles are closed. The abdomen works like a bellows. As it expands and contracts, the valves of the spiracles take turns opening and clos-

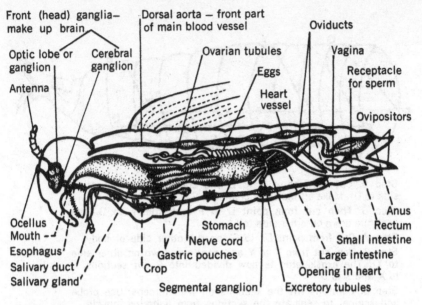

Figure 31. Internal Anatomy of the Female Grasshopper.

ing. Thus fresh air moves in and used air moves out. The faster the grasshopper moves the faster the air circulates through the body.

With Fig. 32 as a guide, use a good light and a hand lens or dissecting microscope to locate the main parts of the respiratory sys

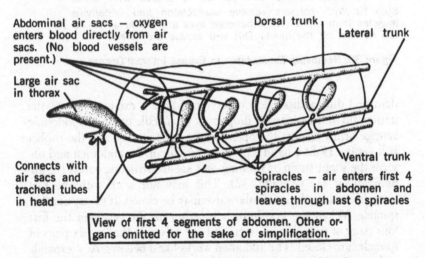

Figure 32. System of Air Tubes (Tracheal Tubes) in the Grasshopper.

tem on your specimen. With scissors, cut a small section of a *tracheal* (air) tube ⅛ inch (3.2 mm) long and examine under a low-power microscope (50× to 100×). Notice the spiral rings that hold the tubes open at all times. The oxygen taken in through the spiracles quickly reaches all the cells in the body. Farmers take advantage of this type of respiratory system in the grasshopper by using *aerosol* (air-sprayed) insecticides. The insecticides are taken in quickly through the spiracles and the tracheal tubes. Thus the poisons used to kill grasshoppers work rapidly.

THE DIGESTIVE SYSTEM

A study of the bottom of Fig. 31 will show the arrangement of the parts of the digestive system. Notice that all the labels related to the digestive system are indicated by broken lines to help you identify the parts in your specimen. The digestive system of the grasshopper is somewhat like that of the crayfish and the earthworm. The digestive organs are easy to find (see Fig. 31). However, it may be necessary, if the specimen is a female, to move the eggs aside in order to see the entire stomach, small intestine, and rectum. Locate the following digestive organs in the specimen:

Mouth—located behind the mandibles. Use your hand lens and a probe to find it.

Salivary glands—located on each side ventrally in the thorax. They send their secretion of saliva into the mouth region through the salivary ducts.

Gullet—a short tube leading from the mouth to the crop.

Crop—a large, thin-walled storage organ that connects with the stomach.

Gastric pouches—several large digestive glands that secrete digestive juices into the stomach.

Stomach—a large chamber in which food is digested.

Large intestine—a short, wide tube that connects the stomach with the small intestine. It conducts wastes into the small intestine.

Small intestine—a short, narrow, coiled tube that carries wastes into the rectum.

Rectum—a chamber shaped like an inflated football that stores wastes temporarily and eventually forces the wastes out through the anus.

Anus—opening at the end of the digestive tract to the outside of the body.

THE EXCRETORY SYSTEM

There are a number of *tubules* at the juncture between the stomach and large intestine. The outer end of each tubule opens into the body cavity and takes in liquid wastes. The wastes are carried by the tubules into the large intestine and ultimately out through the anus.

THE REPRODUCTIVE SYSTEM

Fig. 31 shows the internal anatomy of the female grasshopper. The anatomies of the male and the female grasshoppers are essentially the same. The main external differences are that the female has a pointed abdomen, forked at the tip, while the male's abdomen is rounded at the tip and not forked (see Figs. 27 and 29). Internally, the reproductive organs are different.

In the female grasshopper, a pair of ovarian tubules (small, coiled tubes), located above the gastric pouches, produce eggs which are arranged dorsally like a row of pennies on end. These tubules connect with a pair of oviducts that separate to branch over the large intestine. The branches meet below the rectum and below the nerve cord to form a canal called the *vagina*. Just above the vagina is a small sac, a sperm receptacle, which stores the sperm cells after mating until the eggs are laid, at which time the eggs and sperm meet. The grasshopper uses its ovipositors (see Fig. 31) to force its abdomen into the earth where it forms a burrow. The fertilized eggs are neatly laid in the burrow. There they develop into the young grasshoppers called *nymphs*. When the young grasshoppers are sufficiently developed they emerge from their burrows and start foraging for their meals.

The male grasshopper has two *testes*, which produce sperm cells. The sperm leave each testis through a tube called the *vas deferens*. The tubes leading from the testes unite to form a duct into which glands secrete a fluid. The sperm cells swim in this fluid during mating with the female. An extension of the duct formed by the testes transfers the sperm from the male into the female. If you want to locate these parts, dissect the male grasshopper, using the same dissecting technique as you used for the female grasshopper.

THE NERVOUS SYSTEM

The nervous system of the grasshopper is quite similar to the nervous system of the crayfish. The heads of each have large *ganglia*, which may be called the brain. In the grasshopper, as in the earthworm, the ganglia are ventrally located, segmentally arranged, and connected by a double nerve cord. This is very clearly seen in the grasshopper. Using fine scissors, expose the brain by cutting away the muscles and connective tissue that surround it. Locate the nerve trunk and several of the segmental ganglia.

Now that we have explored the anatomy of the preserved grasshopper, let's try some challenging experiments with live grasshoppers.

Project 1: Fountain of Youth

Why does an insect larva like the caterpillar grow to its full size before it begins the remarkable changes that result in the adult butterfly or moth? One important reason has already been discovered. Part of the larva's brain produces a chemical known as juvenile hormone. As long as this hormone is produced, the larva remains a larva. It does not undergo change or metamorphosis until the hormone secretion stops. Then the larva develops into the adult.

How would you like to try to produce a giant larva or a giant adult insect? It has been done! Several scientists have succeeded in producing giant moths. Let's see if the same experiment will produce giant grasshoppers. To try this you will need the following:

• Several live, young grasshoppers. You can keep them in jars filled with twigs and leaves, covered with perforated screw jar covers. The young grasshopper, unlike the young moth or butterfly, looks like the adult.

• Five to ten abdomens of adult male moths. Moths may be used even if they are old and dried up, because the hormone in the abdomen remains unspoiled indefinitely.

• A mortar and pestle for grinding the abdomens, a tablespoon of clean building sand (obtainable at hardware stores) to mix with the abdomens as you grind them, and about 25 ml of diethyl ether to mix with the ground-up abdomens. *This chemical should be used in school under the supervision of a science teacher. It must be used in a well-ventilated room, away from an open flame.*

• A centrifuge and two centrifuge tubes. A centrifuge is a machine that whirls two or more special centrifuge test tubes around at high speed. When a mixture in these tubes is centrifuged, the heavier materials in the mixture go to the bottom of the tubes and the lighter materials rise toward the top. This helps us to separate the constituents of a mixture. *This too should be done only in school under the supervision of a science teacher.*

• A small glass funnel, round filter paper, two or three glass vials with screw tops (20 to 30 ml in size), 8 ounces (about 0.23 kg) of nonabsorbent cotton, a 50-ml beaker, a glass stirring rod, and a fine artist's paintbrush.

Begin your experiment by placing the moth abdomens in the mortar. Add ½ teaspoon of sand and grind the abdomens with the pestle until the mixture is like a powder. Now continue as follows:

Step 1. Transfer the mixture to the beaker and add enough ether to reach a level of about 1 inch (2.5 cm) in the beaker. Stir with the glass rod, gently but thoroughly; then pour equal amounts of the mixture into two centrifuge tubes to a level of ½ inch (1.8 cm) from the top of each tube. Now centrifuge the tubes for 10 minutes.

Step 2. Fold filter paper as shown in Fig. 33, and place into mouth of funnel as shown. Press filter paper against moistened funnel and insert tube of funnel into one of the vials.

Step 3. Pour liquid from centrifuge tubes on filter paper in funnel. Wait until all the liquid comes through the filter paper into the vial. Stopper the vial with cotton, firmly but not tightly.

Step 4. Place stoppered vial near open window to hasten evaporation of ether. The hormone will appear as a golden-colored liquid.

Step 5. Dip the paintbrush into the hormone and paint a thin stripe on the abdomen of one of the young grasshoppers, but do not cover the spiracles. Keep an untreated grasshopper in a separate container as a control, for comparison. Add several leaves to each container. Keep accurate records of size and other visible changes. Now wait for results! Try this same experiment with other insects. Sooner or later the results will prove exciting.

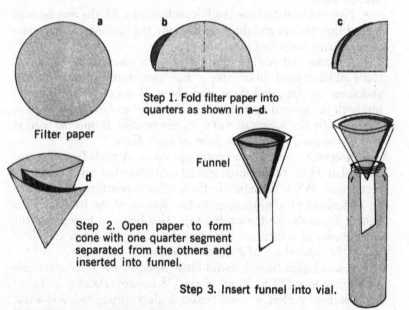

Step 1. Fold filter paper into quarters as shown in a–d.

Filter paper

Funnel

Step 2. Open paper to form cone with one quarter segment separated from the others and inserted into funnel.

Step 3. Insert funnel into vial.

Figure 33. How to Make a Filter.

48

Project 2: How Powerful Are a Grasshopper's Leg Muscles?

Recent scientific experiments have shown that a grasshopper can jump twenty times its own length. If a human 5 feet (1.5 meters, or 1.5m) tall could do that he could jump a distance of 100 feet (30 m), or half an average city block. The grasshopper can jump ten times its body length straight up in the air. Imagine a boy 5 feet tall jumping 50 feet (15 m) high, the height of a four- or five-story building! Clearly, the grasshopper has powerful leg muscles. Can we measure this power? Yes. We can use simplified techniques similar to those used by G. Hoyle, a zoologist, in his studies of the grasshopper. He used electric stimulation on the *anal cerci* of the grasshopper to cause it to raise its legs (see Fig. 34). Weights were tied to the legs and he was able to measure the weight that the grasshopper's legs could lift.

Step 1. Remove insulation from ends of 5 pieces of bell wire and connect to batteries and inductorium as shown.

Step 2. Place a 4" x ½" (10 cm x 1.25 cm) square of plasticine (modeling clay) on a block of wood the same size. With a pencil, press down and form a groove in the plasticine about the size and depth of a grasshopper.

Step 3. Place live grasshopper in groove, turned on its back, its legs sticking up. Gently work the plasticine over its body with your fingers until only the head, legs and the tip of the abdomen along the spiracles remain exposed as shown.

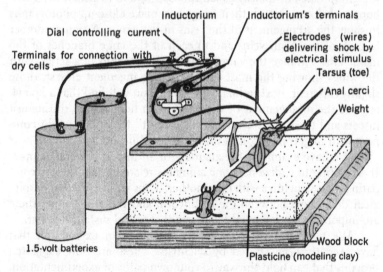

Figure 34. Setup for Measuring Lifting Power of a Grasshopper's Legs.

Let us try to measure the power of a grasshopper's legs with the use of an *inductorium*, a device consisting of electromagnets that is capable of producing small electric shocks when connected to one 1.5-volt or two 1.5-volt dry-cell batteries (see Fig. 34). Inexpensive inductoriums are sold by scientific supply houses.

On each side of the abdominal tip of the grasshopper are tiny structures called *anal cerci*. The anal cerci are sense organs that transmit impulses along the nerve fibers to the nerve centers that control the leg muscles of the grasshopper. The brain is also stimulated by the transmission of impulses from the anal cerci, and relays impulses back to the legs. When these cerci are touched several times quickly with the wire ends of the inductorium the grasshopper responds by flexing its legs to leap. It is the strength of the flexing that we will measure.

Using cotton thread, tie a standard weight of 0.1 gram (abbreviated g) to the *tarsus* or toe of the grasshopper. Then touch the cerci rapidly with the exposed ends of the wires from the inductorium. The legs will flex and raise the weight. Repeat the process with heavier weights, using a different grasshopper for each test, since the tested grasshopper may be injured during the test. Determine finally how much weight a grasshopper will lift.

Weigh the grasshopper on a balance scale in your school laboratory. Compare its weight with the weight its leg can lift. How much weight can the average man lift in comparison with his own weight?

Project 3: Veins Will Tell

Collect different species of grasshoppers and other members of the group (such as locust) called the *Orthoptera* to which the grasshoppers belong. Detach their wings and make close-up photographs to show the arrangement of the veins in the wings. Assign a number to each of the main veins and to each of the large branches of the main veins. Arrange the wings and the photographs in a series, with the wings showing the most similar vein arrangement closest, those slightly different, next in line, and so on. You will find that a knowledge of the arrangement of wing veins can help you to distinguish insects and their relatives. Your "vein print" technique might compare with fingerprinting as a means of identification.

We have examined the anatomy of an insect, the grasshopper. It is exciting to realize how one animal is related to another by comparing their structures with one another. Thus we can see that despite great differences between the earthworm and the grasshopper there are important fundamental likenesses that reveal distant descent.

We have also investigated possible lines of exploration that have recently been started by scientists. These are only suggested beacons that can light the way to your own paths of experimentation. Now let us move on to a different kind of animal, the clam.

5

THE CLAM: A MOLLUSC

THERE IS A LEGEND IN THE SOUTH PACIFIC THAT THE FIRST WOMAN on earth originated within the shell of a giant clam. The belief is that she started a fire in the shell and the smoke attracted an island bird that flew to the clam, settled on the hinge, and pecked at it, striving to open it. When the shell opened, out stepped the first woman! The legend adds that her son became the first chieftain of the island. Yes, there is a giant clam, *Tridacna gigas*. In fact, a giant clam was found that weighed over 500 pounds (227 kg), with a shell about 4 feet (122 cm) long.

No group of animals can outshine the molluscs in brilliance of color, dazzling color combinations, fantastic shapes and patterns. The molluscs outperform most other groups in numbers of different species, ubiquity of dispersal, and survivability. From the ends of the earth to the equator, from the lowest depths to mountainous heights, members of this remarkable phylum can be found.

The molluscs are a venerable group, beginning more than 600 million years ago; from a progressively evolutional point of view, a very live dead-end offshoot of the evolutionary tree. From escargot and clam chowder to shipworms and royal purple dyes, to the poisonous bite of an irritated octopus and the lustrous beauty of pearls, the molluscs are a significant influence on civilization.

The writer–humorist James Thurber said of the oyster, a mollusc: "It has no mind to be burdened by doubt—no fingers to work to the bone. It can never say 'My feet are killing me.' It hears no evil, sees no television, speaks no folly. It produces a highly lustrous concretion of great price, or priceless when a morbid condition obtains in its anatomy."

This is what the oyster, a mollusc, is not! What, then, *is* a mollusc? It is an animal belonging to the phylum *Mollusca*, whose members have a soft,unsegmented body lined by a mantle of epithelium. the outer lining secretes a calcareous, chalklike shell. Molluscs have a ventral muscular "foot," dorsal visceral organs, and gills.

51

The phylum is divided into five classes:

Class 1. Amphineura: Example; the chiton. Shells heavily ribbed

Class 2. Gastropoda: Shells chiefly spirally coiled. Examples; snails and sea slugs

Class 3. Scaphopoda: Shells forming a cone, the mouth bearing filamentous tentacles. Examples; tooth shells

Class 4. Pelecypoda: Bivalves (two-hinged shells). "Foot" hatchet-shaped, as the word pelecypod implies. Large mantle cavity. Examples; clams, oysters, scallops, teredos (shipworms). No structural head

Class 5. Cephalopoda: Foot modified with arms and suckers, pronounced head with strong jaws and rasping radula for abrading food. Examples; nautilus, squid, octopus. (See Chapter 6, "The Squid".)

PREPARING TO DISSECT THE CLAM

The specimens most often used for these dissections are the *Venus clam,* commonly called *quahog;* the *mussel, Mya;* various oysters; and the *Anodonta,* a freshwater clam. The diagrams in this chapter are representative of a freshwater clam and a marine clam, and are intended to be helpful for dissection of either one.

You will need the same dissection equipment used for most dissections (see Fig. 1, p. 5). A good hand magnifying lens of at least 5× would be very helpful. In addition, a strong light as from a microscope lamp is desirable, to illuminate the specimen. As always, work in a well-ventilated room. Wash the specimen with water to remove excess preservative.

EXTERNAL ANATOMY

Step 1. Hold the specimen with the *umbo* above, positioned so that the umbo is to the left of center of the shell (see Fig. 35). The umbo

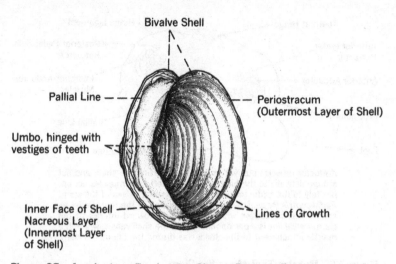

Bivalve Shell

Pallial Line

Umbo, hinged with
vestiges of teeth

Periostracum
(Outermost Layer of Shell)

Inner Face of Shell
Nacreous Layer
(Innermost Layer
of Shell)

Lines of Growth

Figure 35. *Anodonta*, **a Freshwater Clam—External Features.**

is the place where the two halves of the shell are hinged. Each half of the shell is called a *valve*. Your specimen is a *bivalve*, since it has two valves. Examine the outer surface of the valves to see the lines of growth. These lines result from periodic secretions of the organic and mineral matter that produces new shell growth. Unlike tree rings, the lines of growth on shells are not adequately accurate indications of annual growth.

Locate two openings on the right-hand side of the shell, the posterior side. The upper aperture is the terminus of the *dorsal siphon*, also called the *excurrent siphon*, from which water is expelled from the clam. The lower aperture is an intake opening of the *ventral siphon* or *incurrent siphon*. *Cilia*, hairlike extensions of lining cells of the siphons, beat to create the inflow of water bearing oxygen and food. Wastes are also drawn in along with the food and oxygen.

If your specimen has had the two valves separated and free of each other, proceed to Step 3. Otherwise, follow the directions in Step 2.

Step 2. With a sharp scalpel, carefully start to cut at one end of the umbo between the two edges of the valves. Continue to cut, keeping the blade close to the inner face of the upper shell until you have made the incision from umbo back to umbo. Insert a wedge to hold the valves apart. Cut deeper and gently pry the valves until you can insert the scalpel to sever the *adductor muscles* (see Fig. 36). Lift the upper valve and detach it from the lower valve, twisting it to free the tooth lock, and sever the *hinge ligament* at the umbo.

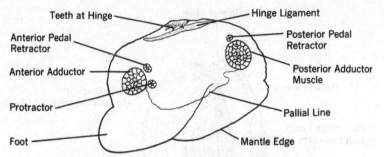

Teeth at Hinge
Hinge Ligament
Anterior Pedal Retractor
Posterior Pedal Retractor
Anterior Adductor
Posterior Adductor Muscle
Protractor
Pallial Line
Foot
Mantle Edge

Adductor muscles pull valves together, closing the shell, and act antagonistically to the hinge ligament. *Retractor muscles* act oppositely to the adductors, serving to open the valves of the clam. *Pedal protractor muscle* produces extension of the foot. *Pedal retractor muscle* causes withdrawal of foot toward interior of the clam. *Pallial line* is a permanent line on the shell valves, where the mantle attachment to the shells was during the life of the clam.

Figure 36. Muscles That Move the Clam.

Step 3. Examine the inner face of the separated valve. Identify the scars of the anterior and posterior adductor muscles, the anterior and posterior *pedal retractor muscles,* and the *protractor muscles.* Locate the *pallial line,* the mark (a curved line) left when the mantle is detached from its mooring on the inner face of the shell. The *mantle* has muscle fibers as well as other types of tissues.

Step 4. Examine the outer face of the shell and the inner face of the freed valve with a hand lens. The tough skinlike outer covering is called the *periostracum* (see Fig. 35). It is organic in composition and protects the middle and inner shell layers from acids in the water, such as carbonic acid, which forms when carbon dioxide is released in water.

The inner layer of the shell is the *nacreous layer.* It is smooth and iridescent, with the pleasing texture and soft glow of a pearl. In fact, the *nacre* is often called *mother-of-pearl.* The middle layer, mostly composed of crystalline calcium carbonate, is called the *prismatic layer* and forms the bulk and weight of the shell. the simplest way to see the prismatic layer is to break a piece of shell and examine the edges with a lens.

INTERNAL ANATOMY—REVEALING THE INTERNAL ORGANS

Lift and turn back the freed *mantle lobe* to expose the *foot,* a large muscular organ used for locomotion, the *labial palps,* small flaps near the mouth that guide the food, and the gills, also called *ctenidia.*

Find the pallial line and the siphons, dorsal and ventral. The siphons are actually mantle tunnels formed by mantle edges that recurve and fuse to form water-conducting channels. They serve the gills, gonads, and kidneys in their respective functions, which will be touched upon later in the chapter.

The structures most readily seen when the mantle, which covers and lines the internal organs, is removed are the viscera and body or *mantle cavity*. In addition, the brownish *digestive gland* around the stomach, the dark overlapping kidneys or *nephridia* that lie below the *pericardium*, the sac that contains the heart, the large foot, and the gills become more readily observable. Study Fig. 37 to determine the relative positions of these structures within the shell.

Carefully cut away the exposed part of the mantle, which you previously freed from the upper valve, at the pallial line. The coiled intestine and rectum can now be observed, as well as the mouth, anteriorly, and the anus, posteriorly. Also locate the pericardial sac, the kidney, and the gonad (see Fig. 37). Then lift the outer gill plate and find the inner gill plate. Cut away (excise) the inner gill plate where it is connected to the foot and save the inner gill plate for further examination. The *genital aperture* (gonadal aperture) opens into the mantle cavity close to the kidney and will now be visible.

Cut through the lower part of the pericardium at its posterior and through the auricles, to expose the *nephridiopores*. With for-

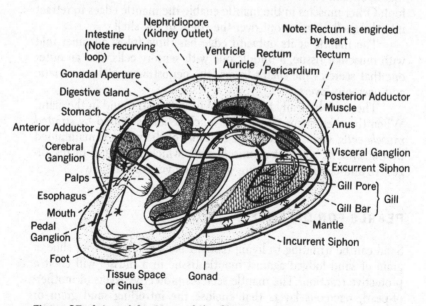

Figure 37. Internal Anatomy of the Clam.

ceps, lift the dissected pericardial flap and fold it forward. The *reno-pericardial openings* will be seen, as will that part of the rectum that passes through the pericardial sac. Use a hand lens to magnify the openings and a probe to find them.

Except for the nervous system and most of the circulatory system, we have located and identified the major parts of the clam's macroanatomy, the organs of the body. We shall now discuss the organization and functioning of its organ systems as we continue to examine the anatomy of the clam.

HOW THE CLAM MOVES ABOUT— MANTLE AND FOOT

The clam's movements may be roughly divided into movements to obtain food and water, and movements to avoid environmental hazards, that is, escape reactions. In either case, muscles are involved. These may cause the clam's shell to close or open (see Figs. 36 and 38a). Other muscles may cause the clam's foot to thrust forward and downward into mud or sand. In addition, there are muscles that move food and wastes along in the digestive system and in the circulatory system.

Where are the muscles located? Fig. 36 focuses on the muscles that open and close the valves of the shell and those that move the foot. Other muscles in the mantle enable the mantle edges to retract into the shell or move out over the edges of the shell.

The mantle at its outward edge has three folds, an inner fold with muscular tissue, a middle one with sensory cells, and an outer one that secretes the shell's layers (the periostracum), the prismatic and nacreous layers.

There are basically two types of muscle cells found in the clam. When the shell snaps closed, the quick action is caused by *striated muscle cells*, while the continued closure of the valves is maintained by the *smooth, unstriated muscle cells*, which are found in the digestive and circulatory systems and in the viscera generally.

PEARLS FOR DEFENSE

Sand can be irritating to living tissue. Like a pebble in your shoe, a grain of sand lodged against mantle tissue in a mussel will elicit a protective reaction. The mantle secretes protective layers of mother-of-pearl, nacreous layers that enclose the intruding sand grain or other foreign agent, and form a pearl. Most fine pearls are formed by

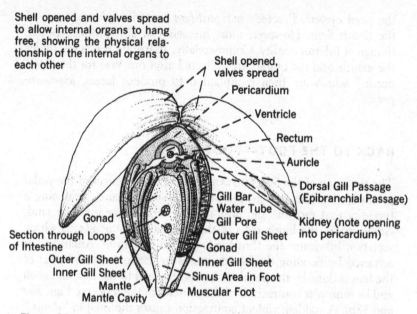

Shell opened and valves spread to allow internal organs to hang free, showing the physical relationship of the internal organs to each other

Shell opened, valves spread

Pericardium

Ventricle

Rectum

Auricle

Dorsal Gill Passage (Epibranchial Passage)

Gill Bar
Water Tube
Gill Pore
Outer Gill Sheet
Gonad
Inner Gill Sheet
Sinus Area in Foot
Muscular Foot

Gonad

Section through Loops of Intestine

Outer Gill Sheet
Inner Gill Sheet
Mantle
Mantle Cavity

Kidney (note opening into pericardium)

Figure 38a. Cross Section of a Clam (Through Heart, Gonad, Foot, and Gills).

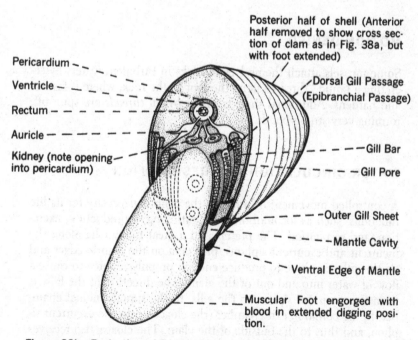

Posterior half of shell (Anterior half removed to show cross section of clam as in Fig. 38a, but with foot extended)

Pericardium

Ventricle

Rectum

Auricle

Kidney (note opening into pericardium)

Dorsal Gill Passage (Epibranchial Passage)

Gill Bar

Gill Pore

Outer Gill Sheet

Mantle Cavity

Ventral Edge of Mantle

Muscular Foot engorged with blood in extended digging position.

Figure 38b. Projection of Foot in Digging Position (see Fig. 38a).

the pearl oysters, *Pinctada margaritifera* and *Pinctada mertensi*, of the South Seas. However, most mussels can also produce pearls, though of inferior quality. Commercially, an irritant is planted under the mantle and the oysters are collected after one year for the "seed pearls," which are then reimplanted to produce larger, jewel-size pearls.

BACK TO THE FOOT—FOOT NOTES

The thrusting of the foot into sand is a combined effect of the pedal protractor muscles and an increase in blood pressure, producing a turgidity that provides the firmness needed to plow into the sand. Additionally, the lining cells of the foot have *mucus glands* whose secretion lubricates the thrust into the sand or mud. Anchoring is achieved by flooding the blood sinuses in the foot. Withdrawal of the foot is done by the pedal retractors attached to the foot and shell and by muscular contraction of the muscles in the foot (see Figs. 38a and 38b). A sudden, violent contraction causes the shell to "jump." Some bivalves, such as scallops, produce motion by jet propulsion, a rapid ejection of water through the excurrent siphon.

MOTION DENIED

Some mussels attach themselves to rocks in turbulent waters by secreting a liquid "thread" along the foot to the rock, where it fastens and hardens. Additional threads called *byssal threads* are spun off, forming very strong bonds.

WATER CIRCULATION AND RESPIRATION

A controlled movement of water in the clam is necessary for its life functions, such as locomotion, food gathering, reproduction, excretion, and respiration. The presence of a great many cilia along the incurrent and excurrent siphons, plus cilia on the mantle edges and on the gills, combine to produce enough propulsive force to cause a flow of water into and out of the clam. The direction of the flow is from the incurrent siphon to the gills, to the suprabranchial chamber, to the common exit chamber (the cloaca), into the excurrent siphon, and thus to the outside of the clam. The cloaca also receives water containing excretory and respiratory wastes.

THE GILLS—LIVING SIEVES

There are two pairs of gills, one pair on each side of the body. Each gill is made up of a dense row of U-shaped structures called *lamellae* (see Figs. 38b and 39). At intervals there are partitions called *interlamellar junctions* penetrated by blood vessels called *branchial vessels.* Inside the lamellae are pores, or *ostia,* for passage of water. Outside are cilia along the ridges and, together, they form the *gill filaments.* The outer part of the outer gill is attached to the mantle. The inner part or lamella of the outer gill, as well as the whole inner gill, is attached to the visceral mass above. This produces a broad channel called the *suprabranchial chamber* or *epibranchial passage.* This chamber receives water from the *water tubes* of the gills. The gill lamellae can move like an accordion, together and apart. They can also shorten and lengthen. That, together with the action of cilia, results in promoting the passage of water through the gills. *Cleansing cilia* on the gill, foot, and mantle surfaces serve to move the heavier sediment from the gills to the mantle edges where mantle muscles contract periodically to expel the sediment. There is, of course, much sediment since the clam lives amid mud and sand. One clam has developed such long siphons that there is not enough room

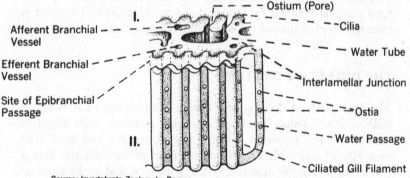

I. View from above, looking down into the gill.
II. View from a side. Lateral view of gill.
Drawn so that I may be set directly over II to form the top of gill.

Ostium (Pore)

I.

Afferent Branchial Vessel

Cilia

Water Tube

Efferent Branchial Vessel

Interlamellar Junction

Site of Epibranchial Passage

Ostia

II.

Water Passage

Ciliated Gill Filament

Source: *Invertebrate Zoology* by Barnes

The open space above the gill is a channel called the epibranchial passage. It is a water passage leading to the excurrent Siphon. The interlamellar junctions are partitions, with ostia and cilia spaced at intervals between successive gill filaments. They contain the branchial blood vessels.

Figure 39. Two Views of a Gill.

in the shell to retract the siphons. This species of clam can be found buried with its siphons sticking out, like a submarine's periscope protruding above the surface of the water.

THE OXYGEN–CARBON DIOXIDE EXCHANGE

The water that passes through the gills at the site of the branchial vessels yields its dissolved oxygen to the vessels, and there the blood is oxygenated. Also, at the branchial vessels the carbon dioxide formed as a result of oxidation in the clam's cells is released to the water that is en route to the excurrent siphon.

EXAMINING A GILL

Study Fig. 39 and review Fig. 37. Place the section of a gill that you had previously cut away under a stereomicroscope, if available, or under a magnifying lens. Observe the U-shaped lamellae and the interlamellar partitions. Find the ostia and scrape a small piece of external "skin" from the lamellae. Place the specimen on a depression slide and add a drop of water and a drop of 0.2% aqueous methylene blue or 0.5% Trypan blue. If neither stain is available, add a drop of household iodine.

Look for tufts of cilia, a number of very thin hairlike structures that protrude at intervals from the outer cells. Finally, note the *food groove* at the base of the lamellae. The cilia above the groove propel food particles trapped by mucus down to the groove. The food is then transported toward the mouth by ciliary action.

THE DIGESTIVE SYSTEM

The main parts of the digestive system from mouth to anus are as follows: labial palps, mouth, esophagus, stomach with digestive gland or "liver" around it, coiled intestine, rectum, and anus. (Review Fig. 37 and study Fig. 40.) Other parts of the clam also have a role in supplying the digestive system with food and in the elimination of food wastes.

We noted previously that cilia in the incurrent siphon, on the gills, and on the mantle edges create a current that moves water into the mantle cavity. The water contains *phytoplankton* and *zooplankton* and their products. The *plankton* are the small plant and animal organisms that live at or near the surface of marine and freshwater bodies.

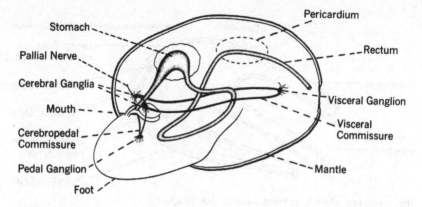

Figure 40. Nervous System and Digestive System of the Clam.

Cilia beat directionally, causing the food to reach the *labial palps*. There is a pair of palps on each side of the mouth. The palps are ciliated and aid in the transfer of food to the mouth. Food reaches the mouth in tiny mucus-covered pellets that are first formed on the gills as the finer particles are sorted out from the heavier sediment. The latter is moved or drops to the mantle edges, where it is expelled. The finer food particles are mired in mucus, which is moved along by ciliary action until directed to the mouth.

MOUTH TO STOMACH

The entire digestive tract has ciliated areas. The cilia and muscular action move the food into a short tube called the esophagus. From there the food reaches the stomach, a large saccular structure (see Fig. 41). Inside the stomach there are folds that guide food to the digestive gland, which almost surrounds the stomach. The digestive gland has many tubules or ducts. The dorsal part of the stomach receives the food from the esophagus. Part of the stomach has a tough lining of *chitin*, the chemical that provides structural strength to insect wings and invertebrate exoskeletons in general.

One region of the stomach is folded and ciliated and is connected with ducts from the digestive gland. Projecting forward from the apex of the stomach is a tough shield of chitin that protects the stomach wall against the rotating style. It is called the *gastric shield* (see Fig. 41). The food is moved in the stomach from dorsal to ventral. Some food moves through digestive gland ducts into the digestive gland. The rest of the food is moved and pressed into a blind pouch in the stomach and thickened with mucus, forming a com-

61

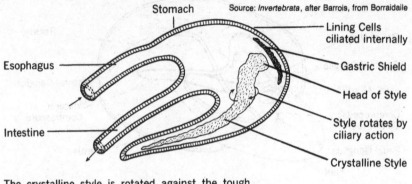

Stomach

Esophagus

Intestine

Lining Cells ciliated internally

Gastric Shield

Head of Style

Style rotates by ciliary action

Crystalline Style

The crystalline style is rotated against the tough gastric shield. As it is abraded the enzyme amylase is released, beginning the digestion of starch.

Figure 41. Section of Stomach of a Clam.

pact, rod-shaped structure called a *crystalline style.* Cilia rotate the style, pressing it like a drill against the toughened gastric shield. In the *Anodonta,* a freshwater clam, the style is believed to rotate eleven times per minute. As the style rotates it releases the enzyme *amylase,* which starts starch digestion. The rotation also results in breaking up larger food masses into finer ones, preparatory to digestion. As the style is rotated it becomes smaller, and the stomach-lining cells secrete replacement material.

THE DIGESTIVE GLAND

The digestive gland consists of a large number of tubules lined with epithelial cells. The gland receives food from the stomach through tubules that connect with the stomach. The food is engulfed by individual cells of the epithelial lining, in a manner similar to the process of *phagocytosis* by human white blood corpuscles. The actual digestion takes place intracellularly. No digestive enzymes are released outside of the cells.

The Giant Clam has an interesting adaptation of phagocytic feeding. Unlike other clams it rests upside down, with its umbo on the bottom. The mantle edges overlap the edges of the valves and algae grow on the overlapping mantle. The algae can receive filtered sunlight because they face the source of light and so they manufacture food by photosynthesis. The clam provides an anchorage and some carbon dioxide needed for photosynthesis in the algae. In turn, the clam harvests the algae, ingesting the algal cells by phagocytosis, and intracellularly digesting the food. Thus the Giant Clam actually maintains a farm on its mantle "fields."

After the food is digested in the cells of the digestive gland, the soluble nutrients are absorbed into the bloodstream. The undigested material is conducted to the small intestine by ciliary and muscular action and ultimately passes through the rectum, anus, cloaca, and excurrent siphon. The rectum has a basal groove that has been compared to the typhlosole of the earthworm. Its function is not clearly defined.

FURTHER DISSECTION OF THE DIGESTIVE SYSTEM

Locate the parts of the system, using Fig. 37 as a guide. With lens and probe, examine the labial palps to see the connection with the mouth. Make a ventral, longitudinal slit in the stomach. With small scissors, cut two small vertical slits to create a flap of the stomach wall. Lift the flap with forceps and locate the stomach folds, gastric shield, and remains of the crystalline style.

Cut into the digestive gland to see its tubular construction. Examine with hand lens of stereomicroscope. Digestion takes place in the lining cells of the tubules.

THE CIRCULATORY SYSTEM— AN "OPEN" SYSTEM

The basic plan of circulation is similar in most clams. There are variations, of course, but all are alike in having an *"open" system*, sometimes called a *lacunar system*. This differs from the "closed" type we find in man, where blood flows from arteries to minute vessels called capillaries, and thence to the veins. Here, in the freshwater clam Anodonta and in all other clams, the blood flows into *tissue spaces* or *blood sinuses* (lacunae), often quite large, instead of through thin-walled capillaries, before returning toward the heart by way of veins.

Study Figs. 42, 43, and 44 for a guide to the circuits found in the Anodonta. Note that there is a three-chambered heart: one muscular ventricle dorsal to two thin-walled auricles, laterally attached to the ventricle. The plan of blood flow is diagrammed in Figs. 42 and 44. Note that there are essentially three main circuits from ventricle back to ventricle. One circuit arises from the *anterior aorta* and goes to the mantle via the *pallial artery*. Oxygenated blood is returned to

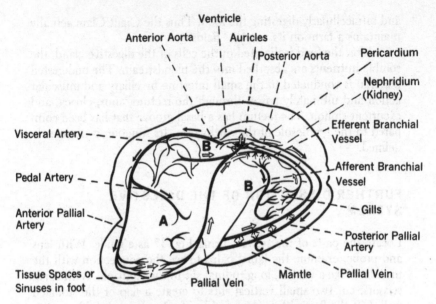

Ventricle

Anterior Aorta   Auricles

Posterior Aorta   Pericardium

Nephridium
(Kidney)

Efferent Branchial
Vessel

Visceral Artery

Afferent Branchial
Vessel

Pedal Artery

Gills

Anterior Pallial
Artery

Posterior Pallial
Artery

Tissue Spaces or   Mantle   Pallial Vein
Sinuses in foot   Pallial Vein

A, B, and C in the illustration indicate the three main
blood circuits starting at the ventricle and returning
blood to the auricles before reentering the ventricle.
Circuit A starts with the anterior aorta and delivers
blood to the mantle.
Circuit B starts with the anterior aorta and delivers
blood to the foot and viscera.
Circuit C starts with the posterior aorta and delivers
blood to the mantle and rectum.

Figure 42. Circulatory System of the Clam.

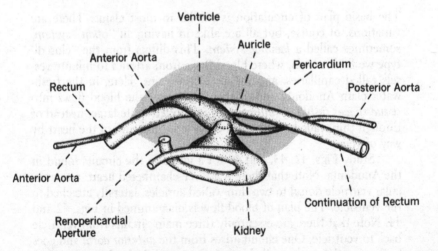

Ventricle

Anterior Aorta

Auricle

Pericardium

Rectum

Posterior Aorta

Anterior Aorta

Continuation of Rectum

Renopericardial
Aperture   Kidney

Figure 43. Heart, Rectum, and Kidney.

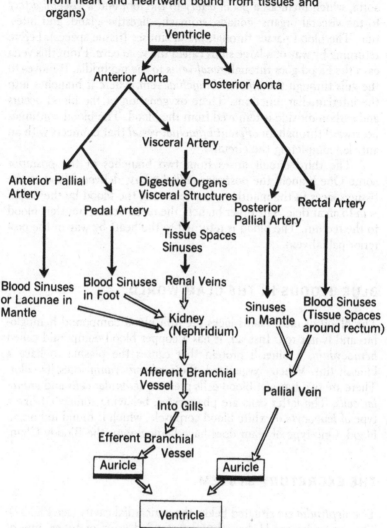

Schematic to emphasize chief circuits (The circuits from ventricle back to ventricle: Outward bound from heart and Heartward bound from tissues and organs)

Ventricle

Anterior Aorta   Posterior Aorta

Visceral Artery

Anterior Pallial Artery   Digestive Organs Visceral Structures   Posterior Pallial Artery   Rectal Artery

Pedal Artery

Tissue Spaces Sinuses

Blood Sinuses or Lacunae in Mantle   Blood Sinuses in Foot   Renal Veins   Sinuses in Mantle   Blood Sinuses (Tissue Spaces around rectum)

Kidney (Nephridium)

Afferent Branchial Vessel

Into Gills   Pallial Vein

Efferent Branchial Vessel

Auricle   **Auricle**

Ventricle

Note that blood from the foot, mantle, and digestive organs passes through the kidney before going on to the gills for oxygenation and back to the heart. Some oxygenation does take place in the mantle where blood is in close contact with environmental water containing dissolved oxygen. That blood bypasses the kidney and returns directly to the heart.

Figure 44. Main Pathways of Blood in the Clam, *Anodonta*.

65

an auricle by a *pallial vein*. Another circuit arises from the anterior aorta, which includes a *pedal artery* to the foot and a *visceral artery* to the visceral organs: gonads, stomach, digestive gland, and intestine. The blood passes through blood sinuses (tissue spaces), before returning by way of a large sinus called the *vena cava*. From this vena cava the blood goes through *renal veins* to the nephridia. It moves to the gills through the *afferent branchial vein*, where it branches into the interlamellar junctions. There oxygenation of the blood occurs and carbon dioxide is removed from the blood. The blood continues heartward through an *efferent branchial vessel* that connects with an auricle, completing the circuit.

The third circuit arises from two branches of the *posterior aorta*. One branch, the *posterior pallial artery*, delivers blood to the tissue spaces in the mantle before returning the blood by the *pallial vein* to an auricle. The other branch, the rectal artery, transfers blood to the rectum. The blood is returned to the heart by way of the posterior pallial vein.

BLUE BLOODS IN THE CLAM WORLD

The blood of most clams lacks the iron blood compound hemoglobin and is not red. Instead, it has a copper blood compound called *hemocyanin*, a blueish protein that causes the plasma to have a blueish tint. When oxygenated, the *oxyhemocyanin* loses its color. There are two types of blood cells present: *agranular cells* and *granular cells*. The latter cells are phagocytes, behaving somewhat like a type of *leucocyte* or white blood corpuscle, which is found in human blood. One type of clam does have hemoglobin: the Bloody Clam.

THE EXCRETORY SYSTEM

Two *nephridia* are situated below the pericardial cavity (see Fig. 37). The nephridia are U-shaped with extended arms or tubes, one of which is coiled above the other. The ventral part opens into the anterior part of the pericardium through an aperture, the *nephridiopore*. The ventral part is considered to be the *kidney*, structurally, whereas the dorsal part is the *bladder* section, which leads into the *suprabranchial chamber*, above the inner gill.

The excretory products enter the pericardial sac through the nephridiopores, one from each nephridium. Filtration of these products then takes place through the heart. The filtered products probably return to the nephridia for removal through the suprabranchial chamber and finally, out of the excurrent siphon.

From the kidney some excretory compounds enter the afferent branchial vein and, in the gills, move into the efferent branchial vein. The efferent vessel joins a *longitudinal vessel* at the kidney and the blood containing the excretory compounds enters the heart for final disposition.

No actual dissection need be done to locate and identify the parts of the excretory system. All you need are a good light, a magnifying lens, and a probe.

THE NERVOUS SYSTEM—CHANNELS OF COMMUNICATION

A nervous system may be called "simple," but that is only in relation to other nervous systems that are more complex. In truth, there is no simple nervous system because nerve cells are highly organized structures with complicated biochemical reactions that are far from being fully understood. Nevertheless, we shall say that "the clam has a relatively simple nervous system."

The clam has three pairs of *nerve centers or ganglia.* One pair, the *cerebral ganglia* (orange colored) is situated near the mouth, one ganglion on each side of the mouth and interconnected by a nerve, the *cerebral commissure.* These ganglia control the anterior part of the body through neural connections (see Fig. 40).

Leading from each of the cerebral ganglia are *cerebropedal commissures,* nerve trunks, that terminate at the two adjacent *cerebropedal ganglia* in the foot. These ganglia and their associative nerves influence the behavior of that area of the body. It is interesting to review the nervous system of the earthworm at this point to see how the cerebral and subpharyngeal ganglia in the earthworm compare with the cerebropedal ganglionic arrangement in the clam (see Fig. 15, p. 20). The *pedal ganglia* with their associated nerves control sense organs or *receptors* in the foot, such as the *statocysts,* which affect equilibrium in the clam.

At the posterior end of the clam are two closely adjacent *visceral ganglia,* lying just below the posterior adductor muscle. The visceral ganglia have trunk-line communication with the cerebral ganglia by means of two *visceral commissures* that pass dorsally between the kidneys to the visceral ganglia. The posterior part of the body is subject to the neural control of the visceral ganglia.

67

RECEPTORS—SENTRIES TO THE CLAM'S WORLD

The clam has several types of *sensory cells* or *receptors*. *Tactile receptors* are sensitive to touch. These are found on the mantle edges and on the siphon closest to the outer environment.

Photoreceptors are sensitive to light. The freshwater clam has no eyes. Some molluscs, such as the scallop, do have eyes along their mantle edge.

Statocysts, receptors that act to maintain balance and equilibrium, are located one on each side of the pedal ganglia. The statocyst is a sac containing fluid and a grain of calcium carbonate, the *statolith*. The grain moves in the fluid, exciting receptor neurons, as the clam shifts its center of balance, delivering its neural message and causing the clam to move back to a balanced position; it acts in effect like a living gyroscope. The same principle operates in the *semicircular canals* of the human ear.

Osphradia are thin patches of yellowish sensory cells on the surface of the visceral ganglia and in the cloaca. They are believed to be involved in detecting and initiating reactions to chemicals, such as smell or possibly taste reactions.

No dissection of the nervous system will be attempted. Instead, study Fig. 40 in relation to your specimen and locate the parts shown in the illustration.

THE REPRODUCTIVE SYSTEM— A SUCCESS STORY

Between the remarkable ways in which the clam and its kin have adapted to their varied environment and their highly effective reproductive "factory," they have produced more than 80,000 different living species of molluscs throughout the world today.

The reproductive mass, the male or female gonad respectively, is a large glandular mass of branching tubules lying athwart the coiled intestine. The *ovary* (*female gonad*) and the *testis* (*male gonad*) each open to the suprabranchial chamber of the inner gill, next to the external pore, the nephridiopore of the bladder (see Figs. 37, 42, and 43).

Egg cells are produced by special lining cells in the ovary and are attached by a stalk to the epithelium. When ready, the *ova* (egg cells) pass out of the ovary through ciliated ducts to the suprabranchial chamber. At spawning time the male sheds its sperm into the water. The female's incurrent siphon sweeps the sperm into its mantle cavity and through the gills' ostia and water tubes, up to the suprabranchial chamber. There eggs and sperm meet and fertilization takes place. The *zygotes* (fertilized eggs) develop in the water tubes of the inner gills and become small bivalves called *glochidia* (see Fig. 45). When small, the glochidia feed on gill secretions until they leave by way of the excurrent siphon to the outside world.

To die or not to die: That is the stark reality facing the glochidium, which resembles a small clam with a bysuss instead of a foot. The glochidium drops to the bottom with its valves gaping, and must hitch on to a passing fish, attaching itself to a fin or gill for food and shelter, until figuratively it can say, "Today I am a clam." When that moment arrives, it gives up its parasitic life and behaves like all its kith and kin, living independently. However, if it fails to snap its valves closed onto a passing fish, its sentence is death.

While on the fish, the glochidium becomes enclosed in a cyst in which it changes into an independent clam. However, it is not truly a mature clam for several years, when its reproductive system finally starts to produce sex cells. The freshwater clam, Anodonta, lives for about 15 years. The odds against a zygote becoming a living adult clam are astronomical. That so many are successful is testimony to the fantastically huge number of zygotes and glochidia that the clam's reproductive factory produces.

Study Fig. 37. In lieu of dissection of the reproductive system,

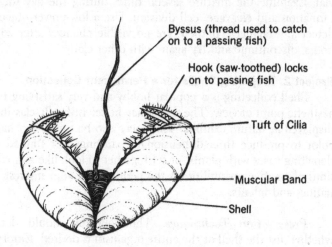

Byssus (thread used to catch on to a passing fish)

Hook (saw-toothed) locks on to passing fish

Tuft of Cilia

Muscular Band

Shell

Glochidia attach to the gills or epidermis of a fish by means of the hook and remain until they metamorphose into young clams.

Figure 45. The Glochidium: A Larval Stage of the Clam.

identify on your specimen the various macroscopic parts of the system: the gonadal mass, its aperture, and its spatial relationship with the other parts of the clam's anatomy.

Project 1: Test-Tube Zygotes

Obtain living mussels from marine or freshwater sources. Open the valves and dissect a small section of the gonad. Place the gonadal tissue in a deep depression slide and add water, preferably from the place where the mussels were obtained. Now tease the tissue with two dissecting needles, using the points to separate and fragment the tissue. Add a cover slip and examine under a microscope; low power first and high power later. It will be necessary to transfer a specimen from the deep depression slide to a shallow depression slide to view with the high-power objective.

Determine the sex of the gametes if your clam was sexually mature and gametes are present. Spring is a likely time to collect gametes. Male gametes, sperm cells, will be very small; female gametes, ova, considerably larger. Mature ova, when present, can be found above an inner gill in the suprabranchial channel.

Remove fluid in the channel with a fairly large pipette or with a large medicine dropper. Add to a depression slide and examine to find ova. If present, add several to a vial or test tube containing water from the clam's habitat. Then, with a pipette, draw up a small volume of male gonadal tissue and transfer the gonadal tissue to the vial. Examine the mixture several times during the day for zygote formation and cleavage, cell division. Use a low-power objective to detect any changes. If there are no visible changes after 24 to 36 hours, discontinue and try again with other clams.

Project 2: Preparing Shells for a Permanent Collection

Shell collecting is a popular hobby and very satisfying from an aesthetic point of view. The shells may be scientifically classified and displayed as nature exhibits. They may also be sorted by shape and color to produce three-dimensional "paintings" or framed murals depicting vases with plants, or set together to look like animals. The limit of creative possibilities is the perimeter of your interest, imagination, and talents.

Preservation Techniques. The collector should determine whether just the shell or the entire organism is desired. Empty shells found along the shores or banks of their native habitat need only be washed and brushed with a clean paint brush.

Shells still inhabited need to have the body removed. This is most easily done by placing the shell in a saturated solution of epsom salts ($MgSO_4$) obtainable in any pharmacy. The amount of solution

70

you need will depend upon the size of the shell. Start with enough water to immerse the shell and add crystals of epsom salts, stir, and continue to add crystals until they can no longer dissolve in the water. Then plunge the shell into the solution and wait about half an hour. Wash the shell with tap water and clean the insides of the shell with a scalpel and small spoon. Dispose of the body, remove the shell, wash it thoroughly, and let it dry.

Preserving the Entire Organism—Shell and All. The organism should be treated with an anaesthetic to relax its muscles and leave it in a normal, uncontracted state. If it is a very large conch shell, immerse it in fresh water and add a saturated solution of epsom salts to the water until the muscles relax. Test by touching the foot or mantle with a probe. If no withdrawal takes place, proceed to the next step, adding the preservatives. Use formalin, obtainable in a drug store. This is a 38% to 40% solution of formaldehyde. You will need to convert it into a 10% solution by mixing one part formalin with nine parts water.

Immerse the anaesthetized and probably dead specimen into the 10% formalin for 48 to 72 hours. Then drain the formalin, which is reusable, and place the specimen in a clean glass container. Add 70% ethanol (ethyl alcohol). Stopper the container tightly. Label the outside of the container. Such a specimen can last a very long time.

Project 3: Studying the Heartbeat of a Mussel and a Snail

Obtain a live clam. Crack open but do not smash one of the shells. A small hammer will do. Using a scalpel, cut through the muscles that hold the shell to the body. Remove the white sheet of mantle tissue from the internal organs. The heart will be seen pulsating. Bathe the exposed area with seawater or artificial seawater to keep the specimen from drying out. (See Project 1, Chapter 7, "The Starfish," for information on artificial seawater.)

You can test the effects of common household drugs like aspirin, Bufferin, etc., on the heartbeat of the clam. Just grind and dissolve a pill in about half a glass of water. Place a drop of the solution over the heart and await the results. Try experimenting with such chemicals as glutathione, cortisone, magnesium sulfate, adrenalin, and insulin.

The snail, *Lymnaea stagnalis,* can also be used for observing heart action. Only young snails will show the heartbeat satisfactorily. The number of bands on the shell of a snail reveals its age; the fewer bands on the shell of a snail, the younger the snail. Try boring or filing a hole in the top of the snail's shell and flashing a pen light into the opening, to get better observations.

71

THE DIGGER AND THE DARTER:
CLAM AND SQUID

The variety of living things is amazing as we view broadly the whole panorama of life. It is also astonishing to find that the quantity and quality of variation is remarkable in members of the same phylum and indeed among members of the same class, genus, and species. Witness the startling differences among the classes of the molluscs, particularly in the next chapter, which deals with the squid.

The clam, in effect, is structurally headless, essentially sedentary, and encased in a hard, hinged shell. The squid of the molluscan class of cephalopods has a well-developed head region, is a swift and strong swimmer, and has only an internal pen, a modified shell, that serves as an internal support structure. Let us move on to this hunter that is, itself, the hunted prey of many of its marine neighbors.

6

The Hunter and the Hunted

THE SQUID

THE SEA HAS SPAWNED AND NURTURED MANY STRANGE CREATURES, from invisible forms to leviathans of the sea, from plankton to Moby Dick. Among the most unusual and interesting of these is the squid (Fig. 46). For centuries the name squid has meant many things to many people. The fisherman thinks of squid as bait; others see squid as food. Many sailors have dreaded the squid as the legendary Giant Sea-Devil or Haf-gufa. Imagine yourself in a longboat idly gazing at the ocean swells, perhaps even trailing your hand over the gunwhales, when suddenly the water parts, revealing "what looks at first like a number of islands surrounded by something that floats and fluctuates like a seaweed. At last several bright horns rise as high as the masts of good-sized vessels. It is said that if the arms were to lay hold of the largest man-of-war they would pull it down to the bottom." This description was written in 1751 in the book *Natural History of Norway*. Is it any wonder that the squid excites the imagination?

The giant squid is 10 to 15 feet (3 to 4.5 meters) long and has tentacles 30 to 40 feet (9 to 12 meters) long. It is rarely seen alive because it inhabits deep waters, but whalers have seen it often in the stomachs of killed whales. The digested squid becomes a greenish gristle with a strange fragrance. This is the fabulous ambergris used in making expensive perfumes.

Of course, not all squid are giants. There are different species. The *Loligo pealii* is probably the most common type of squid in the temperate waters of the Atlantic Ocean. Fishermen who have used this type of squid as bait are little impressed with the ghastly hue of the dead squid. But when seen alive, its beauty of form and color is striking. Disturb it and the colors change rapidly, seeming to flow from one shade to another. To escape its enemies, the squid sometimes seems to disappear behind a black cloud which is formed as it discharges the contents of its *ink sac*.

Figure 46. The Squid *(Loligo pealii)*. (Photograph by Carolina Biological Supply Company)

Legend has it that squid are moongazers. They will swim toward the light of the moon reflected by the ocean and often run aground. Perhaps that is only legend. But it is a fact that fishermen often catch squid by lighting fires on a beach or by hanging a lamp on the prow of a skiff and rowing backwards toward land. The squid follow the firelight or lamplight and become stranded at the shore.

To biologists and other scientists the squid is remarkable in other ways. Together with its close relative, the octopus, the squid is among the most specialized of all invertebrate animals. It has no segments and no exoskeleton, yet it is part of the same species that includes clams, snails, and oysters, all of which do have a shell of exoskeleton. In fact, the squid is more unusual because its body plan is so different from other animals. In most animals it is easy to determine what side is dorsal or ventral and what part is anterior or posterior. In the squid these regions are difficult to identify. To understand this, look at a snail, a relative of the squid. Fig. 47 shows the outline of a snail moving up a glass side of an aquarium. Imagine that you could take hold of the snail at point V and point D and just pull the snail from these points in opposite directions in a straight line. The animal would elongate between V and D and become slender. The result would closely resemble the squid in Fig. 47. Actually no one stretched a snail to make a squid. The curious shape, in

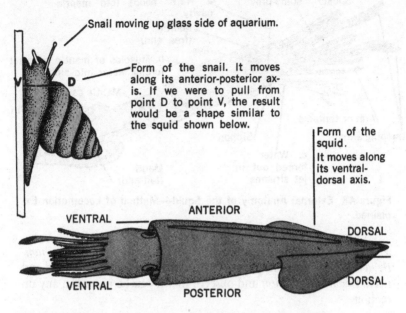

Snail moving up glass side of aquarium.

Form of the snail. It moves along its anterior-posterior axis. If we were to pull from point D to point V, the result would be a shape similar to the squid shown below.

Form of the squid.
It moves along its ventral-dorsal axis.

ANTERIOR

VENTRAL

DORSAL

VENTRAL

DORSAL

POSTERIOR

Figure 47. Comparing Form of a Snail with Form of a Squid.

which the head became the ventral part of the body instead of the anterior part, as in most animals, was evolved by the ancestors of the squid.

The squid may be called the rocket of the sea. Hold it upright and it looks like a rocket resting on its launching platform. It moves by means of jet propulsion, a principle that man has only recently mastered and is now using in missiles, rockets, planes, and small boats. By examining the anatomy of the squid we can understand how it uses the jet stream for locomotion.

EXTERNAL ANATOMY—IDENTIFYING STRUCTURES

To dissect the squid, obtain an injected specimen, the *Loligo pealii*, about 10 inches (25 cm) long. Use the equipment recommended for the dissection of the earthworm.

To study the external anatomy of the squid, place the squid on the dissecting pan. Locate the structures labeled in Fig. 48. Use the hand lens to examine the *suckers* in the tentacles. The suckers are equipped with horny teeth that help to trap the living prey—small fish, crabs, and other squid. Yes, the squid is cannibalistic. Note that two of the arms are longer than the others. These are the *grasping arms* that hold the victim and bring it to the squid's mouth. There is

75

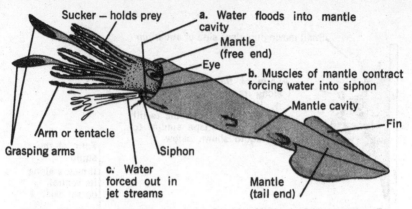

Sucker — holds prey

a. Water floods into mantle cavity

Mantle (free end)

Eye

b. Muscles of mantle contract forcing water into siphon

Mantle cavity

Fin

Arm or tentacle

Grasping arms

Siphon

c. Water forced out in jet streams

Mantle (tail end)

Figure 48. External Anatomy of the Squid—Method of Locomotion Explained.

a loose edge of tissue near the eyes. This is the open end of the *mantle*, a muscular sac that surrounds all of the body except the head and neck. Locate the *siphon* and observe that it can be moved in any direction.

INTERNAL ANATOMY—DISSECTING THE SQUID

Lift the free end of the mantle just above the siphon with a pair of forceps (see Fig. 49). With scissors cut through the mantle in a straight line to the pointed end of the body. Spread the mantle and pin it to the dissecting board as shown in Fig. 50. This exposes the *mantle chamber* and all the internal organs, except those in the head. Locate the supporting cartilages in the free end of the mantle. Trace the siphon backward, using a probe to move aside the muscles that are attached to the siphon and that control it.

Step 1. Lift free end of mantle with forceps.

Step 2. Cut in straight line to tail end of mantle.

Step 3. Spread and roll back cut ends of mantle. Pin to dissecting pan as shown in Fig. 50.

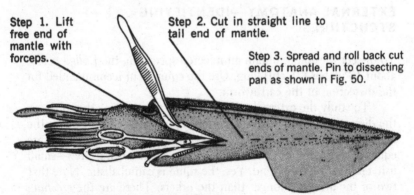

Figure 49. Exposing the Internal Organs.

76

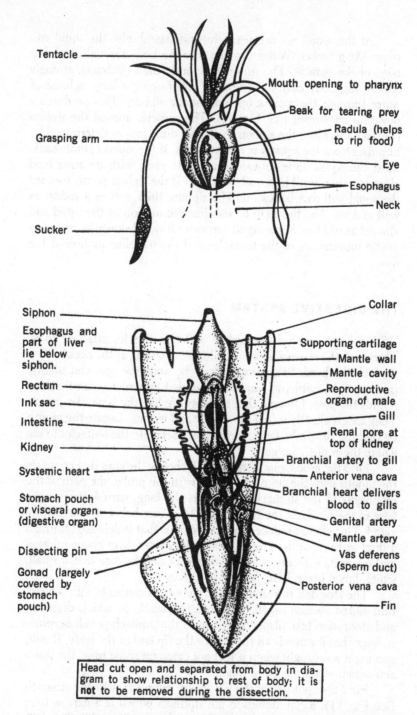

Tentacle

Mouth opening to pharynx

Beak for tearing prey

Radula (helps to rip food)

Grasping arm

Eye

Esophagus

Neck

Sucker

Siphon

Collar

Esophagus and part of liver lie below siphon.

Supporting cartilage

Mantle wall

Mantle cavity

Rectum

Reproductive organ of male

Ink sac

Gill

Intestine

Renal pore at top of kidney

Kidney

Branchial artery to gill

Systemic heart

Anterior vena cava

Stomach pouch or visceral organ (digestive organ)

Branchial heart delivers blood to gills

Genital artery

Mantle artery

Dissecting pin

Vas deferens (sperm duct)

Gonad (largely covered by stomach pouch)

Posterior vena cava

Fin

Head cut open and separated from body in diagram to show relationship to rest of body; it is **not** to be removed during the dissection.

Figure 50. Internal Organs of the Male Squid—Nervous System Omitted.

At this point we are ready to understand why the squid can move like a rocket. Water is taken into the body through the open edge of the mantle. The muscular mantle then contracts, strongly sealing the body cavity at the neck and forcing a large volume of water through the narrow opening of the siphon. This produces a powerful jet stream (see Fig. 48). The muscles around the siphon contract and cause the siphon to change direction, as determined by impulses from the squid's nervous system. If the siphon points backward the squid darts forward toward its prey, with its arms held close together, until it is ready to strike. If the siphon points forward the squid will dart backward. The siphon, then, acts as a rudder as well as a jet. The fins help to stabilize the motion of the squid and also act as rudders. Some squid can swim forward slowly by means of gentle movements of the tentacles and the wavelike motion of the fins.

THE DIGESTIVE SYSTEM

With scissors, cut off the siphon and the muscles attached to it. Then with the scissors make a shallow cut starting at the neck and up through the head, to a point midway between the eyes and slightly past the eyes, until you reach the rounded muscular organ that surrounds the jaws. Observe that the jaws are like the beak of a parrot. They are excellently suited for cutting and tearing. Locate the mouth by pushing the probe between the jaws. Separate the tentacles to see where the probe emerges.

Find the *esophagus*, a narrow tube below the jaws that connects the mouth with the stomach. Trace, with the probe, the path of the esophagus to the stomach as it enters the long, narrow liver at the base of the head. At the front of the liver, just below it, is a salivary gland. This gland produces a digestive juice that is delivered through a duct to the mouth. The esophagus leaves the liver about midway and connects with the stomach after it passes the *pancreas*, a small, white, lobed organ lying below the kidneys.

The stomach is a small, thick sac which connects with a large, thin-walled visceral organ, the *stomach pouch*, in which digestion and absorption take place. After a meal the stomach pouch becomes so large that it extends all the way to the tip end of the body. If your specimen was caught when it had not eaten for some time, the stomach pouch would be quite small.

Find the spot where the esophagus connects with the stomach (see Fig. 51). Right alongside the stomach pouch is a narrow tube that extends toward the head. This is the intestine. The intestine

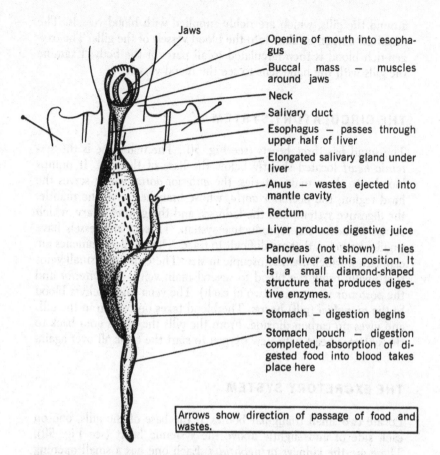

Jaws

Opening of mouth into esophagus

Buccal mass — muscles around jaws

Neck

Salivary duct

Esophagus — passes through upper half of liver

Elongated salivary gland under liver

Anus — wastes ejected into mantle cavity

Rectum

Liver produces digestive juice

Pancreas (not shown) — lies below liver at this position. It is a small diamond-shaped structure that produces digestive enzymes.

Stomach — digestion begins

Stomach pouch — digestion completed, absorption of digested food into blood takes place here

Arrows show direction of passage of food and wastes.

Figure 51. Simplified Digestive System of the Squid.

connects with a small rectum and an opening called the anus. Use the probe to help you find the rectum and anus. Wastes are expelled from the body through the anus.

Above the rectum is the ink sac (see Fig. 50). It opens into the siphon near the anus. When in danger, the squid forces its "ink" into the water and moves away, leaving a dark cloud that conceals it from its enemies.

THE RESPIRATORY SYSTEM

Locate the *gills* (see Fig. 50), one on each side of the body. They look like curved feathers pointing toward the head. Water containing dissolved oxygen is taken into the mantle cavity. The water swirls around the gills, which are richly supplied with blood vessels. The dissolved oxygen is taken into the blood vessels of the gills. The oxygen-rich blood is then circulated to all parts of the body. Examine the gills with the hand lens to see the blood vessels.

THE CIRCULATORY SYSTEM

The squid has three hearts (see Fig. 50). The main one is the *systemic heart* located slightly below the base of the gills. It pumps blood into three main arteries: the *anterior aorta*, which serves the head region; the *posterior aorta*, whose branches serve the mantle, the digestive system, and the kidneys; and the *genital artery*, which carries blood to the reproductive system. These main vessels have smaller branches that are difficult to trace. The smallest branches are capillaries, which are microscopic in size. These run into small veins that finally bring the blood to several main veins, the *anterior* and the *posterior vena cavae* (two of each). The vena cavae deliver blood to the *branchial* (gill) *hearts*. The blood takes on oxygen in the gills and gives up carbon dioxide. From the gills the blood goes back to the systemic heart with new oxygen to start the cycle all over again.

THE EXCRETORY SYSTEM

Locate two small triangular bodies at the base of the gills, one on each side of and slightly above the systemic heart (see Fig. 50). These are the kidneys or *nephridia*. Each one has a small opening into the mantle cavity through which liquid wastes are released. Solid wastes are discharged through the anus (see Fig. 51).

THE NERVOUS SYSTEM

The main part of the nervous system is in the head and neck where it is most needed (see Fig. 52). It is very much like the nervous system of the grasshopper, crayfish, and earthworm, consisting of several main ganglia (nerve centers), connectives, and nerves that go to all parts of the body. In live squid the large *stellate* (star-shaped) *ganglia* are visible through the transparent lining of the mantle where the neck and mantle meet.

Recent investigations have shown that the squid has giant-sized nerve fibers that make excellent subjects for the study of nerve reactions.

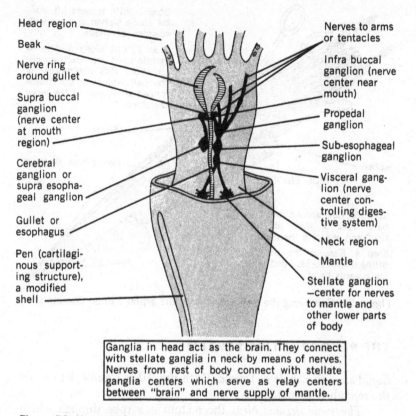

Head region

Beak

Nerve ring
around gullet

Supra buccal
ganglion
(nerve center
at mouth
region)

Cerebral
ganglion or
supra esopha-
geal ganglion

Gullet or
esophagus

Pen (cartilagi-
nous support-
ing structure),
a modified
shell

Nerves to arms
or tentacles

Infra buccal
ganglion (nerve
center near
mouth)

Propedal
ganglion

Sub-esophageal
ganglion

Visceral gang-
lion (nerve
center con-
trolling diges-
tive system)

Neck region

Mantle

Stellate ganglion
—center for nerves
to mantle and
other lower parts
of body

Ganglia in head act as the brain. They connect
with stellate ganglia in neck by means of nerves.
Nerves from rest of body connect with stellate
ganglia centers which serve as relay centers
between "brain" and nerve supply of mantle.

Figure 52. Nervous System (Main Centers) of the Squid.

THE SKELETAL SYSTEM

The squid is an invertebrate with an endoskeleton or inner skeleton. This is most unusual because normally only vertebrates have an endoskeleton. After you have completed the dissection and identified the internal organs, remove the pins that hold the mantle to the dissecting pan. Roll the cut mantle back to its original shape and position. Pin the two loose edges together. Turn the squid over completely so that its anterior (see Fig. 47) surface faces up, and remove the inner shell or *pen* as shown on Fig. 53. You will remember the squid is related to the clam. The pen of the squid is really a shell that has been modified and encased in tissue.

The squid has a casing of cartilage tissue in the head that is somewhat like the cartilage that covers the brain of the shark, which is a vertebrate animal. The squid also has internal cartilage supports in the mantle tissue.

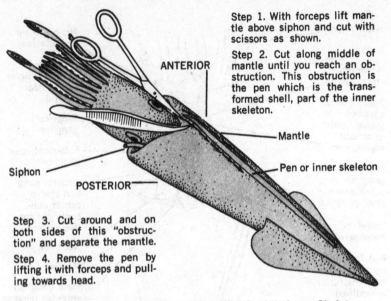

Step 1. With forceps lift mantle above siphon and cut with scissors as shown.

Step 2. Cut along middle of mantle until you reach an obstruction. This obstruction is the pen which is the transformed shell, part of the inner skeleton.

ANTERIOR

Mantle

Pen or inner skeleton

Siphon

POSTERIOR

Step 3. Cut around and on both sides of this "obstruction" and separate the mantle.

Step 4. Remove the pen by lifting it with forceps and pulling towards head.

Figure 53. Removing the Pen—A Part of the Squid's Inner Skeleton.

THE REPRODUCTIVE SYSTEM

Squid are either male or female. With Fig. 50 as a guide, let's study the reproductive system.

The male organs: Near the rectum is a tube through which sperm cells are discharged. This tube connects with a complicated tube called the *vas deferens* that extends down below the gills and there connects with the testis, a large, long gland that produces sperm cells.

The female organs: Alongside the left branchial heart is a swollen sac called the *oviducal gland.* This connects with a long, coiled tube called the *oviduct* that is filled with eggs during the breeding season. If you have a female squid, try cutting the oviduct with a razor and examining the eggs with a low-power microscope (50× to 100×) or with a 5× hand lens. Note the large mass that looks like a cluster of many tiny pearls. The oviduct runs into a large, lobed gland, the *ovary,* which is located above the stomach pouch and extends to the tip of the body.

In our discussion of the squid we learned that the snail and the squid, two very unlike forms (see Fig. 47), belong to the same major group of animals, the *Mollusca.* The following projects will help you expand your knowledge about these animals.

Project 1: Cockles, Mussels, and Devilfish

After dissecting the clam, try dissecting the octopus to help round out the picture of this strange and important group of animals. Follow the same general directions in dissecting the octopus as you did for the squid. Obtain an injected specimen of *Octopus vulgaris*.

For the clam dissection get an injected *Anadontia mutabilis*, a freshwater clam or mussel. Bear in mind that the foot, a fleshy extension between the shells or valves, is comparable to the head and neck of the squid, and that the pen of the squid is actually a reduction or modification of the shell of the clam. The general body plan is comparable in both types.

Project 2: A Living Wood Drill—The Shipworm

The shipworm or *teredo* is not a worm. It is a member of the molluscs with a modified toothed shell. The shell is located at the front of the body and is used by the animal for boring chambers into wood. It destroys wooden hulls of ships, piles, docks, and wharves, particularly in the tropical and subtropical zones.

Obtain a specimen of wood containing live shipworms or *Teredo navalis*. Remove one from its wood chamber and dissect it, using the same general procedure as you used in dissecting the squid. You will see remarkable adaptations for its strange way of life. The February 1961 issue of *Scientific American* has an article on the shipworm which provides excellent illustrations that may be used as a guide for this dissection.

The squid family has become important because of the many research investigations being carried on with different squid. At Woods Hole on Cape Cod, Massachusetts, exciting things have been done with molluscs, under the auspices of the Marine Biological Institute for Research and the Oceanographic Institute.

Let us now look at an important invertebrate whose shape, though attractive, appears to make it an unlikely candidate for being in the least compatible with the development of the chordates. Yet, among the members of the starfish group (the *Echinodermata*) were organisms believed to have provided the genetic revolution linking invertebrates with the once and future vertebrates.

7

Star of the Sea

THE STARFISH

THE STARFISH IS NOT A FISH. IT HAS NO INTERNAL SKELETON LIKE that of a fish or other vertebrates. Why then should we be interested in dissecting the starfish? Scientists believe that some member of the starfish family became the ancestor of the backboned animals, the vertebrates, many millions of years ago. It is fascinating to see how different they are from vertebrate animals. Figs. 54 and 55 show top and bottom views of a starfish.

The starfish is an enemy of man. It can destroy huge beds of oysters, which are an important food source for humans. The starfish eats oysters in a curious way. First it settles its rays, or arms, around

Figure 54. Starfish—Topside.   Figure 55. Underside of the Starfish.
(Photograph by Carolina Biological Supply Company)

the shell of a living oyster. Then it starts a tug-of-war, the oyster using its muscles to keep the shell tightly closed, the starfish trying to force the shell open. The struggle is uneven. The oyster depends entirely on its muscles, which sooner or later tire. The starfish uses the power of the sea (described on page 89) and never tires. As the oyster's shell finally opens, the starfish causes its stomach to move out of its body and around the flesh of the victim. The starfish then secretes digestive juices from the digestive glands in the rays, and dinner is over. The starfish takes in little waste material from the food it eats because the food is digested externally and, as can be expected, it has a poorly developed intestine and rectum.

The most important weapon of a scientist is knowledge. It was knowledge of the structure and habits of the starfish that made it possible to cut down the size of the starfish population and thus save the oyster industry. For many years oystermen used to catch and smash starfish and throw the pieces back into the sea. But they discovered that this practice almost ruined the oyster industry because starfish have the power of regeneration, or regrowth. Many of the rays that were pitched back into the sea grew into whole new starfish, thereby further endangering the oyster beds. Research led to the present methods of using sea mops—mops made of cloth or string that are dragged along the sea bottom to catch the starfish, which are then left in the sun to dry and to die. They can also be killed in hot water tanks, or by exposure to live steam (steam that comes directly from the boiler at its highest peak of temperature).

PREPARING TO DISSECT THE STARFISH

For dissecting purposes, obtain an injected specimen of *Asterias forbesi* or *Asterias rubens*. Use the same basic equipment as you used for the earthworm dissection. In addition, you will need a pair of strong, sharp, pointed scissors about 4 inches (10 cm) long, a hand magnifying lens with a diameter of 1 inch (2.5 cm) or more and about 5× magnification, and a glass microscope slide. A low-power microscope (20× to 50×) would be helpful.

A stereozoom microscope would be very helpful for this type of dissection because close examination is called for. The stereozoom type is especially valuable as greater magnification can be achieved in continuous focus by merely turning a knob. Fig. 56 shows one type of stereomicroscope. Many schools have one or more microscopes available for students and for demonstration purposes.

Fig. 57 will be your guide to dissecting the starfish. Because each arm, or ray, is like the other rays on both the outside and inside, we can dissect one arm to see the digestive organs, another arm to see

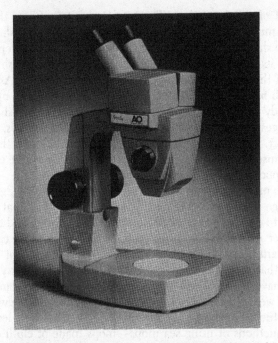

Figure 56. Stereozoom Microscope—A Valuable Instrument for Examining and Dissecting Small Specimens and Small Areas of Larger Specimens. (Courtesy AO Scientific Instruments)

the reproductive organs, and so on. Fig. 57 is arranged in seven parts, marked A to G. Each part shows where to dissect and shows views of the different internal organs. Your dissected starfish should look like Fig. 57 when you have completed the dissection.

EXTERNAL ANATOMY

Before dissecting the starfish, let us examine the external characteristics. Place your specimen in the dissecting pan, dorsal side up. The dorsal side has no grooves or tube feet (see Fig. 54).

Locate the five rays, or arms. (Specimens may have four rays or even fewer, if rays have been broken off prior to handling. In rare instances, starfish have been found with up to twenty-five rays.) Now look for the very small circular disc located near the starfish's center. Note that it is slightly off center. This is the *sieve plate* (see Figs. 54 and 57). The sieve plate is a perforated disc through which water enters the body of the starfish.

86

With the paintbrush used in the grasshopper experiment or with a piece of cheesecloth wrapped around your finger, stroke the skin on each ray. Note its spiny nature. Examine the skin with the hand lens to see the spines in closer detail.

Turn the specimen ventral side up. With Fig. 57 as a guide, examine the tube feet (part D) with the hand lens. Find the mouth in the center of the specimen (see Fig. 55). Notice the groove that runs from the tip of each ray toward the center of the starfish (see Fig. 55). The tube feet are arranged on each side of the groove (part D of Fig. 57).

INTERNAL ANATOMY

Let us begin the dissection as shown in part A. Cut off about 1 inch (2.5 cm) from the end of one ray. Force the sharp, pointed end of the scissors through the skin at the side of the ray. Then cut the skin around the ray in a ring. Move the skin slightly to one side and snip off the end of the ray.

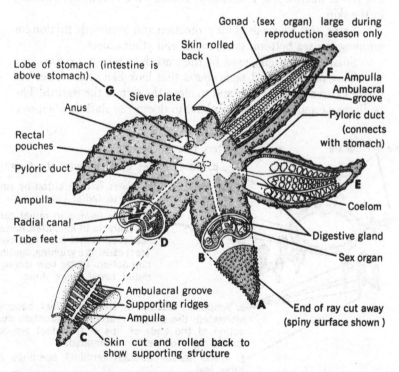

Figure 57. Dissection of the Starfish—External and Internal Anatomy (Dorsal View).

87

Examine the cut end of the ray at part B with your hand lens. Note the feathery-looking digestive glands and, under them, the sex organs near the beginning of the ray where they join the central part of the body.

To understand something about the framework of the ray, cut off the end of another ray as shown in part C. Cut it off in the same way as in part A. Then, using forceps, clean out the digestive organ and sex organs, if present. (If sex organs are not at peak of development they will not extend as far as the last inch of the ray.) Now cut the skin with scissors and roll it back, as shown in part C. Locate the structures labeled in part C. Examine the inside wall of the ray with the hand lens. You will see the supporting ridges, the bulblike *ampullae* (tiny sacs that are part of the tube feet described in Fig. 58), and tiny openings in the inner wall. These openings are pores that connect with the projecting gill tube and are part of the external gills that help the starfish to breathe.

Cut off a strip of skin about ¼ inch (6 mm) wide from the loose flap shown in part C. Place the piece of skin on the glass slide with the spines pointing upward. Examine with the hand lens or with the low-power microscope, if available. Identify the following structures in the skin:

Large, stiff spines used for protection and to provide friction for gripping the sea bottom, sand, rocks, and other objects.

Small, hairlike gills used by the starfish to take in oxygen.

Tiny pincers called *pedicellaria* that look like pliers. They are used to grip minute objects and to clean the skin of the starfish. The pedicellaria and the spines combine to doom the starfish when oys-

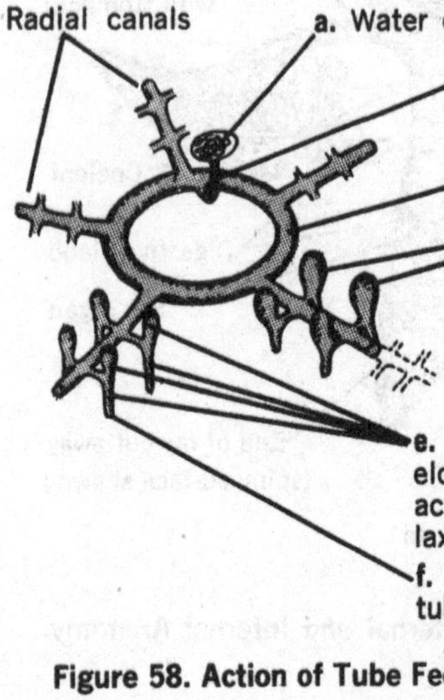

Radial canals

a. Water enters through sieve plate.

b. Water moves down through stone canal.

c. Water is distributed by ring canal to radial canals.

d. Water flowing in radial canals collects in ampullae. Ampullae expand and tube feet contract causing a gripping, suction cup action. Water flow through radial canal slows down.

e. Ampullae reduced in size, tube feet become elongated; the starfish relaxes grip. (Suction cup action of the ends of lips of tube feet are relaxed.) Flow of water accelerated.

f. Water leaves through terminal openings of tube feet.

Figure 58. Action of Tube Feet of the Starfish—Dorsal View.

88

termen use their sea mops to clear the starfish from the oyster beds. These structures attach themselves to the sea mops and thus the starfish are removed from the sea to perish on land.

TRANSPORTATION BY TUBE FEET

To enable you to identify better the water transport, or water vascular system that helps the starfish move about, the digestive gland and the sex organs have been omitted in part D of Fig. 57. Raise the digestive gland of the specimen with a probe or with forceps and find the structures labeled in part D. Count the number of tube feet on 1 inch (about 2.5 cm) of skin. How many inches are there in all five rays of your starfish? Measure the length of one ray. Multiply this length by five, the usual number of rays on the starfish. Now estimate the total number of tube feet on the rays of the starfish by multiplying the number of tube feet on 1 inch (2.5 cm) by the combined length of all the rays. Does this help you to understand one reason why the starfish can force open the shell of a resisting oyster?

The starfish has a simple, primitive nervous system without a brain. But it does have a highly developed method of locomotion resulting from the action of its tube feet, the special structures that are connected to a system of canals (see Fig. 58). Most animals draw their power from the action of their muscles. While the starfish has muscles, its main source of power is the sea. It uses the water of the sea to create a form of suction at the ends of its tube feet. This suction enables the starfish to move and to grasp objects. As some of the tube feet hold on to the sea bottom, others release their hold. The free part of the starfish then contracts by means of its muscles and moves toward the part that is holding to the sea bottom. The rear tube feet then take hold and the front ones are released. In this way the starfish "pulls" itself along. As long as there is water the starfish has power.

THE DIGESTIVE SYSTEM

The digestive system of the starfish is made up of the following parts:

Mouth—a small opening in the center of its lower or ventral side.

Stomach—a lobed, muscular sac shown by the dotted lines in part G. Each lobe, or section, of the stomach has a tube called the *pyloric duct* that connects with the digestive gland in each ray.

To expose the digestive gland and pyloric duct in one ray, cut a section of the skin away as shown on part E. Using scissors, start the cut near the tip of the ray along one side. Then cut in a straight line toward the center of the starfish. Do the same along the other side and cut the skin off where the ray meets the center of the body. The digestive gland is now exposed. Raise the digestive gland gently with a probe and trace its connection to the pyloric duct. Examine the digestive gland with a hand lens to see the many tiny structures that secrete the digestive juice.

Intestine—a thin sac just above the stomach (see part G of Fig. 57) in which some digestion takes place and wastes are moved toward the rectal pouches and the anus.

Rectal pouches—two very small sacs that store small amounts of wastes temporarily. Little is actually known about these structures.

Anus—A very short tube coming from the intestine, and opening to the outside through the upper or dorsal part of the starfish. Wastes are discharged through the anus.

THE REPRODUCTIVE SYSTEM

To study the reproductive organs (part F, Fig. 57) cut open another ray in almost the same way as described in part E of Fig. 57. First lift with forceps the exposed digestive gland and cut the pyloric duct close to the place where it joins the stomach. Now remove the entire digestive gland. This will enable you to more easily expose the *gonads* (reproductive organs), which are below the digestive gland and on each side of the pyloric duct.

During the breeding season the reproductive organs enlarge until they almost fill the entire ray, but at other times they will be small. You will probably get starfish with small gonads because the supply houses generally do not collect starfish during the breeding season. The starfish bodies are so crowded with eggs in the breeding season that other organs are hard to see. Moreover, the starfish are easily damaged at that time and are hard to preserve.

The testes of the male starfish discharge sperm and the ovaries of the female starfish discharge eggs into the water during breeding season. The sperm and eggs join together and the unions develop into bilaterally symmetrical larvae that develop into adult starfish. Although both the testes and the ovaries look alike, page 93 describes how to tell them apart.

LOCATING THE DIGESTIVE SYSTEM

To expose the stomach, intestine, rectal glands, and anus of the starfish, carefully study part G of Fig. 57 before dissecting. (This area is, of course, covered by the starfish's spiny surface.) Note that the dotted lines form the outline of the stomach. Remember that the intestine is *above* the stomach and very close to the skin. With these two facts in mind, cut, with scissors, a circle of skin about 1½ inches (3.8 cm) in diameter from the top center of the starfish and remove it. Make the cut as shallow as you can. Even then, the intestine will probably be damaged. Be careful not to disturb the *sieve plate* and the *stone canal*, the tube that leads down from the sieve plate (see Fig. 58). Locate the stomach and the other labeled structures in that area. Push a probe downward through the stomach. It should pass through the gullet and out through the mouth without meeting any obstruction.

Try to trace the stone canal to its connection with the ring canal (see Fig. 58 for details). Use forceps to pick away tissue from the canals. This operation is difficult. If you succeed in doing this neatly you may have the makings of a fine surgeon!

A HEARTLESS, BLOODLESS CIRCULATORY SYSTEM

The dissection of the starfish has shown you that the starfish is very different from the crayfish or the grasshopper and their cousins, and that the starfish is really a very simple kind of organism as compared with the crayfish and the insects. It has no heart or blood vessels, and circulation is accomplished mostly by water flowing through the open body cavity or *coelom*. Sea water is its blood and the open, continuous spaces in its body substitute for blood vessels. Water enters through the gills in the skin and moves into the body cavity (coelom). There it is forced to circulate by cells with little projections that lash the water as the starfish adventures on the ocean bottom. The starfish has small *ameboid* cells—cells that can change their shape. These behave very much like certain types of white blood corpuscles in our blood.

Yet, in spite of his many anatomical omissions, this interesting little fellow and his family contributed a larval ancestor that gave rise to the highest group of animal life, the vertebrates. Read about the *bipinnaria larva* of the starfish and the theoretical larva, the *dipleurula*. It will help you to understand better the links between the invertebrates and the vertebrates. Compare this larva with *tornaria* larvae of the higher group of animals. The resemblance emphasizes evolutionary linkage.

There are many projects that are begun simply because we become interested in an idea. Often such projects result in the development of very satisfying lifelong hobbies that provide color, interest, drama, and many opportunities for scientific research for science-oriented people. Following are several projects that may enrich your life for many years to come.

Project 1: Ocean Life in Your Home

Many people keep a freshwater aquarium stocked with tropical fish in their living rooms. Maintaining such aquaria is an interesting hobby that serves as a conversation piece and as a source of living beauty. Few people know that a marine aquarium can be maintained in the home and can be even more attractive and exciting than a freshwater aquarium.

You do not have to live near the sea to have a marine aquarium. Small chemical packets may be purchased from biological supply houses that, dissolved in fresh water, produce artificial seawater in which marine life can flourish. Directions for the use of the chemicals and for the care of the aquarium are provided by the suppliers. However, experience has shown that some natural seawater is desirable if the animals are to survive for extended periods of time. A satisfactory ratio is about 1 U.S. gallon (0.833 Imperial gallon or 3.8 liters) of natural seawater to 9 gallons (about 34 liters) of artificial seawater. A 1:10 mixture will do the job as well. There are several things you have to be careful to do when you set up a marine aquarium:

- Use an all-glass tank or a tank where no metal will come in contact with the water. (Metal reacts to salts in "sea" water and produces compounds poisonous to fish.) Handle the animals with a plastic salad spoon and fork combination. Do not touch the animals with anything made of metal.
- Keep the tank away from strong light or direct sunlight, which can be injurious to the fish. Use fluorescent lighting. Hobby shops sell tanks equipped with fluorescent lighting.
- Use clean beach sand only. No other kind of sand is satisfactory. The sand can be purchased if necessary. Include plants like sea lettuce and cladophora. They are good oxygen producers.
- Aerate the water continuously with an aquarium pump which drives air into the water.
- Use two tanks, a large tank (about 10-gallon size) and a small tank (about 2-gallon size). The small tank is used for feeding the animals, and since the fish are kept there for ½ hour or less ordinarily, it does not need any lighting or any of the other trimmings.
- Keep the main tank near a cold window. Keep the window open at the bottom about 2 inches (5 cm) if possible. Try to hold the

temperature of the water below 60 degrees Fahrenheit (16 degrees Celsius). The temperature may be kept down during the warm months by adding to the water in the tank, daily, a tray full of ice cubes from a rubber or plastic tray. (No metal ions will be transferred into the tank if ice from a rubber or plastic tray is used.)

• Do not crowd the tank. About six to ten small animals are all that a 10-gallon tank can maintain. You can use such creatures as starfish, sea urchin, clam, sea snail, barnacle, very small fish, scallop, sponge, and jellyfish.

• Feed the animals sparingly.

It is especially interesting to change the scene every so often by discarding old stock—exchange with friends or provide "decent burial"—and experimenting with different combinations of animals. Note how they react to each other. Some get along with each other, some are indifferent to their neighbors, while others do something about their "grievances."

Project 2: Is It a He or a She?

The male and female sex organs of the starfish look alike. How can we be sure which one is which? This can best be done with fresh starfish which you may be able to collect or get from a fishing boat. If that is not possible for you, then preserved specimens will have to do.

Dissect an arm of the starfish as shown in part F of Fig. 57. Locate the sex organs and remove a tiny piece of tissue from the sex organ with forceps. Place the tissue sample on a microscope slide. Add a drop of water to the tissue. Then spread or tease the tissue with two dissecting needles. Place a cover slip over the thinned out tissue and examine it with a microscope, first under low power and then under high power (400× or higher). The male organs, testes, will show minute cells that have a tiny tail or flagellum. These cells are the sperm cells. The female organs, ovaries, will show cells that look like tiny pearls. These cells are the egg cells. They are much larger than the sperm cells.

8

A Boneless Fish

THE DOGFISH SHARK

THROUGHOUT THE WORLD OF THE SEA THERE ROAMS A POWERFUL creature, the shark, whose very name excites fear and horror. Yet few types of sharks will attack a human without provocation. The dogfish shark (also called a sandshark), like other sharks, is a scavenger, feeding chiefly on dead matter. Most sharks are relatively harmless to man, but all sharks are a source of irritation to the commercial fisherman because they prey upon the fish he wants to catch and so interfere with his livelihood.

A study of the history of living things from prehistoric time shows that sharks have always successfully inhabited the seas. Fossil evidences reveal that the shark was as prominent hundreds of millions of years ago as he is today. While other primitive forms were much modified and became extinct with the climatic changes that have altered the earth's interior and surface structure, the shark lives on as one of the most evolutionarily successful forms of life ever produced. Why is this so?

The shark has been so successful because it is relatively free to roam the seas, for which it is excellently suited by its streamlined form and great strength. Another reason is that sharks protect their young by carrying them in their bodies until the young sharks are almost fully formed. This protection helps the group to survive, since few of the young are destroyed by other animals. The shark's body is a highly muscular machine and its scales form a tough armor. In fact the scales are really primitive teeth, and the fearsome teeth in the mouth of the shark are actually modified scales (see Fig. 59).

Most fish have bones that form an inner skeleton, as many of us have unhappily discovered at the dinner table. However, the sharks and the rays (primitive fish closely related to sharks) do not have bony skeletons. Instead their inner skeleton or *endoskeleton* is com-

94

Rows of sharp tipped scales (enlarged) — tips point backward. Scales are embedded in skin and continue into mouth where they are specialized as teeth. The teeth, like the scales, are superficial and have no roots.

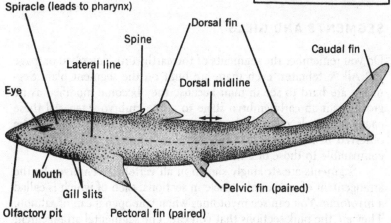

Figure 59. External Anatomy (Side View) of the Dogfish Shark.

posed of a tissue called *cartilage*. While cartilage is softer than bone, it still has the quality of toughness. It has yet another important property: it is somewhat flexible. This permits great freedom of bodily motion. To get some idea of the properties of cartilage, hold the upper part of your ear between two fingers and then twist it. The same kind of tissue that is under the skin of your ear makes up the endoskeleton of both the small dogfish shark and of the giant basking shark, which reaches a length of 40 feet (about 12 meters)! The shark, like the ray, is a primitive fish. The cartilaginous skeleton in the shark is a forerunner of the bony skeleton found in higher types of fish like the herring.

MENU FOR THE SHARK

One of the unceasing challenges facing all animals is the fateful problem of survival—to get food or to become food. Many animals have difficulty in surviving because their diet is very limited. For example, deer are vegetarians; if a prolonged and heavy snowfall blankets their food, they starve. But sharks are omnivorous; that is, they will eat almost anything. In fact, studies of the contents of sharks' stomachs have shown some remarkable things, such as tin cans, indi-

gestible rubbish thrown overboard from boats, and many other unlikely foods. The shark can get rid of such indigestible matter because it has an *evertible digestive organ;* that is, a part of its intestine that can turn itself inside out (see Fig. 53 and page 100: the *spiral
valve*). That is how the shark discharges indigestible objects like fish
bones and other wastes from its body.

SEGMENTS AND GILLS

Do you remember the segments of the earthworm described on page
11? All vertebrates, even man, are built on the segment plan. Segments are hard to see in man because they become modified as he
grows from an early embryo stage to a late embryo stage. All these
changes take place before the human baby is born. Segments may be
observed in the segmented gill arches of a human embryo that are
comparable to those of a fish.

Segments are strikingly shown in all vertebrate embryos by the
arrangement of masses of muscle in sections, each of which is called
a *myotome*. You can see myotomes when you open a can of salmon.
They are the pink sections that you eat. The segmental arrangement
of muscles in the salmon is very much like that in the dogfish shark.
Segments in the dogfish shark show clearly in the gill slits and gill
arches (see Figs. 59 and 67).

Since the dogfish shark lives only in water, we are not surprised
to find that it has no lungs and breathes by means of gills. You may
ask why a whale, which also lives in water, has lungs and no gills.
The answer is that the whale is a mammal, not a fish. Its ancestors
were "landlubbers" with four legs. Like all mammal embryos, including man, the embryo whale has gill slits.

Unlike most fish, the dogfish shark is a live-bearer; that is, it develops its young inside the body instead of discharging its eggs to develop into fish outside the body. The young dogfish shark is often
called a *puppy*. You will find out more about the puppy when you do
your dissection. It is important to study the shark because a dissection of the shark reveals the vertebrate pattern of structure in its
basic primitive form.

DISSECTING THE DOGFISH SHARK

Obtain two specimens of the dogfish shark *Squalus acanthias*, one
male and one female, from 12 to 18 inches (30 to 46 cm) long. If the
female is carrying a developing baby shark you can actually study
three specimens, the male, the female, and the puppy. The male dogfish shark has *claspers*, which he uses to channel or direct the sperm

(male sex cells) into the female's reproductive chamber; the female dogfish shark does not have claspers.

In order to perform the dissection you will need the same equipment you used for the earthworm dissection, a pair of sharp, curved surgical scissors about 6 inches (15 cm) long, a smooth metal probe about 6 inches (15 cm) long, and a large dissecting pan or board about 20 inches (51 cm) long. You will also need a roll of plastic wrap or aluminum foil to wrap the specimens when you are not working with them so that they do not dry out. A magnifying lens 1 inch (2.5 cm) or more in diameter is important to help locate and trace nerves and blood vessels, and a pen flashlight helps to trace the origin of small vessels and to locate small apertures. We are now ready to examine and dissect the dogfish shark.

EXTERNAL ANATOMY

Place the male dogfish shark on the dissecting pan, dorsal side up. (Mouth is on the ventral side.) Locate on the specimen the unpaired fins or single fins on the dorsal midline, the paired pectoral fins, and the single tail or caudal fin (see Fig. 59). Locate all labeled apertures or openings on the dorsal and ventral surfaces of the head and on the ventral side between the pelvic fins (see Fig. 60). Examples of apertures are spiracles, mouth, gill slits, nostrils, and cloaca.

Turn the dogfish ventral side up (mouth upward) to examine the under part of the specimen. Identify on the specimen the follow-

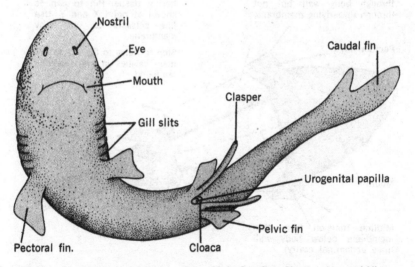

Figure 60. External Anatomy of the Male Dogfish Shark—Ventral View.

ing parts labeled in Fig. 60: pair of pectoral fins, pair of pelvic fins, pair of claspers, and caudal fin. Uneven lobes in the caudal fin are characteristic of all sharks.

OPENING THE ABDOMINAL CAVITY

The specimen that you receive has a deep transverse or crosswise cut in the tail region. Cut the tail piece off at this point with a scalpel. Then dissect and peel back about 2 inches (5 cm) of the skin on the cut-off tail piece. Use probe to help separate the skin from the muscles. You will see the segmental arrangement of muscles or myotomes. Save the tail piece for study of a vertebra or spinal bone. (See Project 3, Chapter 9.) Now follow directions for opening the abdominal cavity as given in Fig. 61.

BODY CAVITIES

All animals from the earthworm to man have an internal cavity called the *coelom*. Consult Fig. 62 to see the two main body cavities in which the major organs are found.

Step 1. Place specimen in pan ventral side up and pin pectoral fins as shown.

Step 2. Using scalpel make midline incision cutting from point A to E and E to B. Cut through body wall but not through underlying membrane.

Step 3. Cut from point A to C on each side and from E to D on each side.

Step 4. Roll back flaps of body wall. Use probe or scalpel to free body wall from any adhering tissue. Pin to pan as shown at point X and Y. Use fine scissors to cut away membrane.

Step 5. Refer to Fig. 62 to see main cavities in which organs are found.

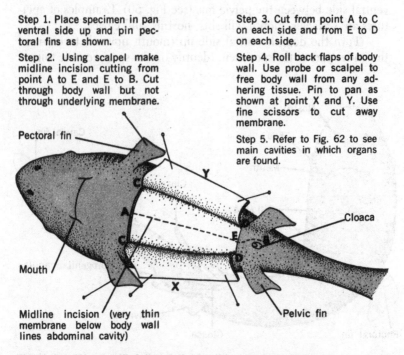

Pectoral fin

Cloaca

Mouth

Midline incision (very thin membrane below body wall lines abdominal cavity)

Pelvic fin

Figure 61. Opening Abdominal Cavity of the Dogfish Shark.

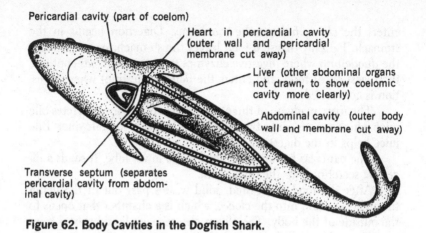

Pericardial cavity (part of coelom)

Heart in pericardial cavity (outer wall and pericardial membrane cut away)

Liver (other abdominal organs not drawn, to show coelomic cavity more clearly)

Abdominal cavity (outer body wall and membrane cut away)

Transverse septum (separates pericardial cavity from abdominal cavity)

Figure 62. Body Cavities in the Dogfish Shark.

THE DIGESTIVE SYSTEM

To understand how the shark's digestive system functions we will follow the digestive organs in the specimen, with the aid of the diagram in Fig. 63 which shows the abdominal organs.

Food entering the *mouth* passes through the *pharynx* and through the *esophagus*. The pharynx and esophagus are not shown in Fig. 63 because they are not found in the abdominal cavity. Food

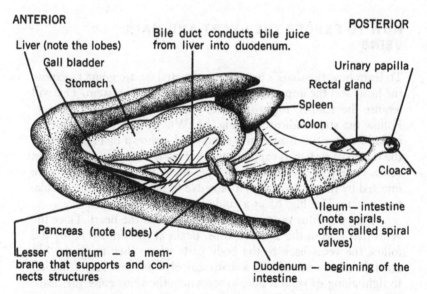

ANTERIOR

POSTERIOR

Bile duct conducts bile juice from liver into duodenum.

Liver (note the lobes)

Gall bladder

Stomach

Urinary papilla

Rectal gland

Spleen

Colon

Cloaca

Ileum — intestine (note spirals, often called spiral valves)

Pancreas (note lobes)

Lesser omentum — a membrane that supports and connects structures

Duodenum — beginning of the intestine

Figure 63. Digestive System (Abdominal Organs) of the Dogfish Shark.

enters the *stomach* from the esophagus. Digestion begins in the stomach. Partially digested food leaves the stomach, passes through the *duodenum* where digestive enzymes from the *pancreas* continue the digestion of food, and enters the *ileum* (intestine) where digestion is completed.

The *liver*, made up of three main divisions or lobes, secretes bile juice which empties into the duodenum through the bile duct. Bile juice helps in the digestion of fatty foods.

The pancreas has an upper lobe and a lower lobe. It sends a digestive secretion into the duodenum.

After the food is digested, solid wastes pass out of the ileum through the *colon* into the *cloaca*, which is a chamber that opens to the outside of the body. The cloaca also receives liquid wastes from the *urinary ducts* which terminate in the *urinary papilla*, and sex cells from the reproductive organs (see Figs. 68 and 69). Solid and liquid wastes are expelled from the cloaca through the cloacal opening.

The *spiral valve* is a muscular sac inside the ileum or intestine which turns itself inside out through the cloacal opening like the waste bag in a vacuum cleaner and empties its store of indigestible wastes into the ocean.

The *spleen*, although one of the abdominal organs, is not considered a part of the digestive system. It is actually a part of the lymphatic system, a division of the circulatory system. Little is known about the physiology of the spleen.

HOW TO EXPOSE THE HEART AND MAIN VEINS

To learn how the shark's heart is constructed we are going to expose the heart and the important veins of the circulatory system. This will require the scalpel, forceps, probe, and a pair of straight scissors. Follow the steps indicated in Fig. 64. What the circulatory system looks like after the dissection is completed is shown by Fig. 65. After the digestive system is exposed, move it aside to locate all the main veins. Note that the veins in the specimen appear blue. They were injected by the supplier with a blue dye to help us trace the circulatory system. Use Fig. 65 as a guide in studying the veins.

Find the *sinus venosus*, an antechamber of the heart. Trace the veins that lead to the sinus venosus. Begin at the sinus venosus and follow the veins back to the body parts where they originate. Slit open the sinus venosus with a sharp scalpel or a razor blade, from left to right along its ventral side, to see where the veins enter the heart. Lift each vein with a fine probe to help you trace its course.

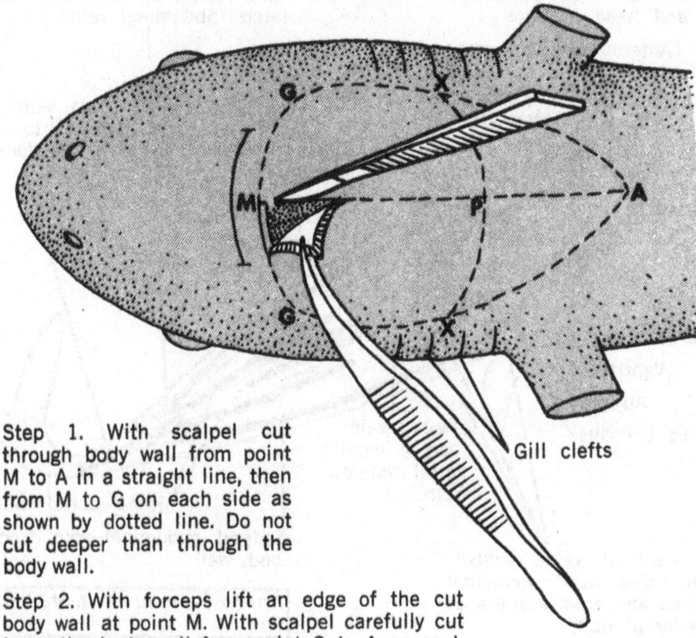

Step 1. With scalpel cut through body wall from point M to A in a straight line, then from M to G on each side as shown by dotted line. Do not cut deeper than through the body wall.

Step 2. With forceps lift an edge of the cut body wall at point M. With scalpel carefully cut loose the body wall from point G to A on each side.

Step 3. With scissors cut off the loosened body wall. Follow the dotted curve MGA on each side of body as you cut.

Step 4. The pericardial sac containing the heart is now exposed. With scalpel cut open this thin-walled sac from point A to P and from P to X on each side.

Step 5. Carefully dissect away muscles in area from M to P, above dotted lines XPX, up to gill clefts on both sides. Avoid cutting large blood vessels and nerves.

Figure 64. How to Expose the Heart and Main Veins.

THE HEPATIC PORTAL SYSTEM

How does the digested food reach all parts of the body of the dogfish shark? This is accomplished as follows: The digested food is transported from the digestive organs (stomach, duodenum, and ilium) through veins of the *hepatic portal system*. (Hepatic means pertaining to the liver.) From the liver the blood containing the digested food flows through hepatic veins into the sinus venosus. The blood then goes from the sinus venosus into the heart, which pumps the blood through the arteries to all parts of the body.

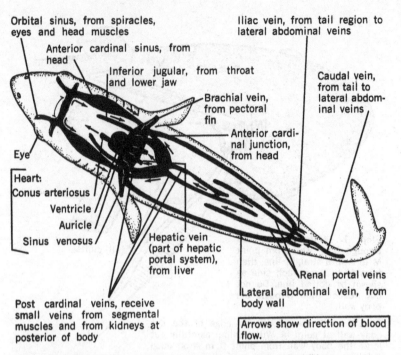

Orbital sinus, from spiracles, eyes and head muscles

Iliac vein, from tail region to lateral abdominal veins

Anterior cardinal sinus, from head

Inferior jugular, from throat and lower jaw

Brachial vein, from pectoral fin

Caudal vein, from tail to lateral abdominal veins

Anterior cardinal junction, from head

Eye

Heart:
Conus arteriosus
Ventricle
Auricle
Sinus venosus

Hepatic vein (part of hepatic portal system), from liver

Renal portal veins

Lateral abdominal vein, from body wall

Post cardinal veins, receive small veins from segmental muscles and from kidneys at posterior of body

Arrows show direction of blood flow.

Figure 65. Main Veins of the Dogfish Shark—Simplified Diagram.

Fig. 66 shows a simplified version of the hepatic portal system emphasizing the main veins. With the aid of the diagram, locate the veins of the hepatic portal system in your specimen.

THE MAIN ARTERIES

We have examined the system of veins in the dogfish shark. Now let us examine the system of arteries, which serves only one purpose, to deliver blood to the organs. To study the arteries of the dogfish shark we are going to continue the dissection, following the procedure described below:

Trace the arteries in the abdominal section from the heart to the main organs they service. Make an incision with the scalpel from one corner of the mouth to the first gill slit on one side of the head. Lift the skin with forceps and separate it from the underlying tissue with a scalpel and probe until the skin is free from the mouth up to the first gill slit. Lift the exposed sections of muscles until the blood vessels around the gills are exposed. Be careful not to cut any large

102

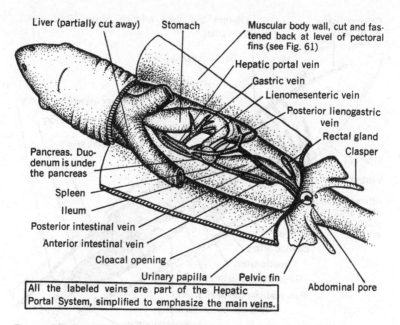

Liver (partially cut away)　　Stomach　　Muscular body wall, cut and fas-
tened back at level of pectoral
fins (see Fig. 61)

Hepatic portal vein

Gastric vein

Lienomesenteric vein

Posterior lienogastric
vein

Rectal gland

Clasper

Pancreas. Duo-
denum is under
the pancreas

Spleen

Ileum

Posterior intestinal vein

Anterior intestinal vein

Cloacal opening

Urinary papilla　　Pelvic fin

Abdominal pore

All the labeled veins are part of the Hepatic
Portal System, simplified to emphasize the main veins.

Figure 66. Simplified Hepatic Portal System of the Dogfish Shark.

nerves or blood vessels. The nerves are white in color. In your specimen the arteries contain a red dye while the veins contain a blue dye.

Follow the arteries with a probe. Use scalpel and probe to free the arteries from any adhering tissue in the anterior end (the head). Using Fig. 67 as a guide, determine which organ is served by each of the main arteries labeled in the illustration.

THE MALE REPRODUCTIVE ORGANS

Fig. 68 illustrates the reproductive organs of the male dogfish shark. To locate these organs in the specimen, lift the liver and move its lobes aside. Do the same thing with the digestive organs. The *urogenital system* (excretory and reproductive organs) will then be seen.

The testes produce sex cells called sperm. The sperm cell has a tail or *flagellum* that enables it to swim from the testes through small tubes into the *Wolffian duct*. In the male dogfish shark the Wolffian duct, in addition to sperm cells, carries urine from the kidneys to the cloaca. In the female the Wolffian duct transports only urine.

From the Wolffian duct of the male the sperm cells enter the *seminal vesicles* and *sperm sacs* for temporary storage. During mat-

103

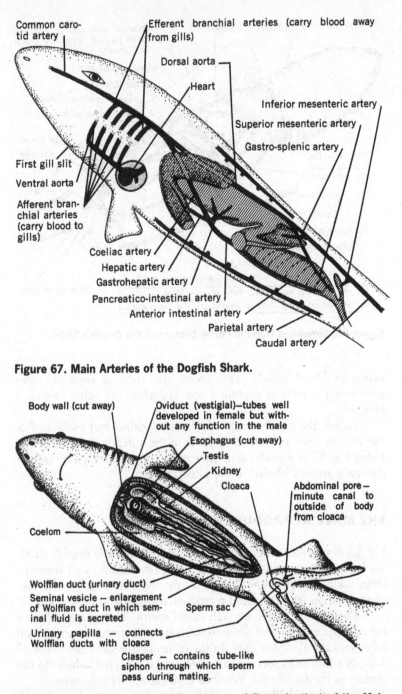

Figure 67. Main Arteries of the Dogfish Shark.

Figure 67 labels:

Common carotid artery

Efferent branchial arteries (carry blood away from gills)

Dorsal aorta

Heart

Inferior mesenteric artery

Superior mesenteric artery

Gastro-splenic artery

First gill slit

Ventral aorta

Afferent branchial arteries (carry blood to gills)

Coeliac artery

Hepatic artery

Gastrohepatic artery

Pancreatico-intestinal artery

Anterior intestinal artery

Parietal artery

Caudal artery

Figure 68 labels:

Body wall (cut away)

Oviduct (vestigial)—tubes well developed in female but without any function in the male

Esophagus (cut away)

Testis

Kidney

Cloaca

Abdominal pore—minute canal to outside of body from cloaca

Coelom

Wolffian duct (urinary duct)

Seminal vesicle — enlargement of Wolffian duct in which seminal fluid is secreted

Sperm sac

Urinary papilla — connects Wolffian ducts with cloaca

Clasper — contains tube-like siphon through which sperm pass during mating.

Figure 68. Urogenital System (Excretory and Reproductive) of the Male Dogfish Shark.

ing the sperm cells are discharged from the sperm sacs through the cloaca, through the *urinary papilla* and along a channel on the claspers to the cloaca of the female dogfish.

Despite the fact that sharks are live-bearers, there are considerable variations in the patterns of growth and development of shark embryos among the different species of sharks. For instance, in the sand tiger shark's uterus a deadly internal struggle for survival takes place. As the zygotes develop into the shark pups they begin to attack and devour each other until a lone survivor remains who is the victor and starts a new generation.

Provision for the developing young takes place in some sharks by the production of very large, yolk-filled eggs, sufficient to nourish the young until birth. One species produces an egg the size of a large grapefruit. In other cases, the developing embryos receive nutrients from the mother by transference from her bloodstream, through capillaries, to the bloodstream of the embryos; this is an early precursor of the highly developed placenta of mammals. (See Fig. 108, p. 169, to find the uterus, site of placental formation.) Another species of shark produces as many as eighty offspring at a time, supplying each developing embryo with a sufficiently large egg to support its full embryonic life, before it emerges to become an independent predator.

THE FEMALE REPRODUCTIVE ORGANS

After the sperm cells of the male dogfish shark unite with the eggs of the female, the fertilized eggs develop within the *uteri* (singular, uterus) of the female's body into young sharks or puppies. The puppies develop a food sac or *yolk sac* on their ventral, abdominal section that provides food for the young sharks until they are discharged from the mother's body to go hunting their own food.

What are the female shark's reproductive organs? The reproductive system of the female dogfish shark is the basic pattern for the reproductive systems of all higher vertebrate animal forms.

To dissect the female dogfish shark, follow the same directions given in Figs. 61 and 64 for the dissection of the male dogfish shark. Move the liver and the digestive organs aside to expose the reproductive organs. Note the absence of claspers in the female dogfish shark. Starting at the ovaries, identify parts of the reproductive system of the female dogfish as labeled in Fig. 69.

Dissection of the female shark is especially interesting if the specimen contains developing young. Cut open each uterus and examine the contents for possible developing embryos or puppies.

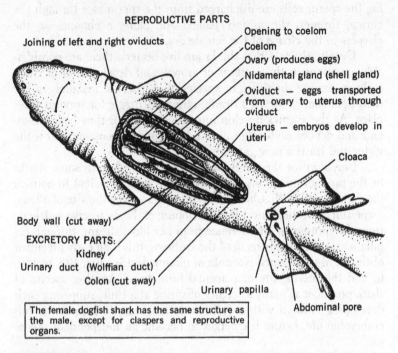

REPRODUCTIVE PARTS

Joining of left and right oviducts

Opening to coelom

Coelom

Ovary (produces eggs)

Nidamental gland (shell gland)

Oviduct — eggs transported from ovary to uterus through oviduct

Uterus — embryos develop in uteri

Cloaca

Body wall (cut away)

EXCRETORY PARTS:
Kidney
Urinary duct (Wolffian duct)
Colon (cut away)

Urinary papilla

Abdominal pore

The female dogfish shark has the same structure as the male, except for claspers and reproductive organs.

Figure 69. Urogenital System (Excretory and Reproductive) of the Female Dogfish Shark.

THE NERVOUS SYSTEM

Can the shark think? Does it have a high IQ? When we examine the nervous system of the dogfish shark (see Fig. 70), we find that it definitely has a brain which is far more advanced than that of the invertebrates we have studied. However, it is primitive and poorly developed compared to the brain in the higher vertebrates. The shark's brain is very much like the human brain in elementary pattern, but the main parts are not developed to the same degree as in the human brain. For instance, the cerebrum, or thinking and association area, is greatly developed in humans, but in the dogfish shark it is small and poorly developed. On the other hand, the olfactory area that controls the sense of smell is very large in the shark's brain but is quite small in the human brain. The dogfish shark has ten pairs of cranial nerves (see Fig. 70) that control the sensory and motor activities of the head and some other parts of the body. Man has twelve pairs of cranial nerves that perform similar functions. These examples are enough to show some of the principal changes that took place in the evolution of the nervous system as more advanced vertebrate animals appeared on the earth.

We are going to expose the nervous system of the dogfish shark by following the procedure described below:

Place the dogfish shark on pan dorsal side up. Cut away all the skin from top of head to last gill clefts. Remove the cranium (cartilaginous covering over brain) by slicing away very thin, flat sections of cartilage with a sharp scalpel until the brain is exposed. Be very careful not to cut any nerves, especially near the eyes. The nerves look like thin white cords. Some nerves pass through the cartilage.

Locate the main parts of the brain (see Fig. 70) and the connection of each part with the spinal cord. The *medulla oblongata* is the end of the brain and connects with the spinal cord. The spinal cord has paired spinal nerves.

After dissecting the shark it would be valuable to dissect the

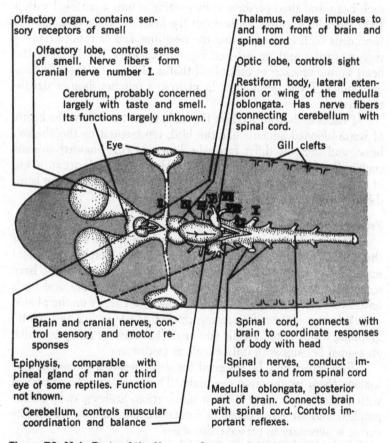

Olfactory organ, contains sensory receptors of smell

Olfactory lobe, controls sense of smell. Nerve fibers form cranial nerve number I.

Cerebrum, probably concerned largely with taste and smell. Its functions largely unknown.

Eye

Thalamus, relays impulses to and from front of brain and spinal cord

Optic lobe, controls sight

Restiform body, lateral extension or wing of the medulla oblongata. Has nerve fibers connecting cerebellum with spinal cord.

Gill clefts

Brain and cranial nerves, control sensory and motor responses

Epiphysis, comparable with pineal gland of man or third eye of some reptiles. Function not known.

Cerebellum, controls muscular coordination and balance

Spinal cord, connects with brain to coordinate responses of body with head

Spinal nerves, conduct impulses to and from spinal cord

Medulla oblongata, posterior part of brain. Connects brain with spinal cord. Controls important reflexes.

Figure 70. Main Parts of the Nervous System of the Dogfish Shark—Dorsal View, Cranium Removed.

perch, a bony fish, to highlight the evolutionary change in endoskeleton from cartilage to bone.

SOME INTERESTING PROJECTS

Did you ever stop to think that there is a connection between the kind of environment animals live in and the type of organs they have? Here is a suggestion for making a comparative study of hearts of various animals to test this relationship.

Project 1: A Comparative Study of Hearts

Make a collection of vertebrate hearts. Begin with the lowest vertebrates like the dogfish shark or other fish. To remove the heart, follow the same general procedures given in Fig. 64. Photograph each heart and then preserve it by putting it into a jar filled with a 10% formalin solution. Dissect out the heart of preserved amphibia specimens such as the frog and the necturus, and of various reptiles such as turtles and snakes (see Fig. 90, Chapter 10). Remove the heart from a chicken or other fowl that is planned for the gala Sunday dinner. Try to get a cow's heart from a butcher shop or slaughterhouse.

Dissect each heart to show the main parts. Show how the hearts of warm-blooded animals like the bird, represented by the chicken heart, and the cow differ from the hearts of cold-blooded animals such as the dogfish shark and the frog. How does the heart structure of amphibia, which live part of their lives in water and part on land, differ from the hearts of land-living vertebrates?

Project 2: Tips from Scales

Each major group of fish has its own kind of scales. The scale of the dogfish shark is called a *placoid* scale. The perch has *ctenoid* scales. Other animals like the reptiles, birds, and mammals also have some kind of scales. Birds' feathers are related to scales and birds have scales on their legs. Even humans have tiny scales on their hairs.

You can make a fascinating project of preserving and photographing scales. After a while you can determine the group to which an animal belongs with only a scale as evidence.

Fish scales may be softened for study under a microscope by soaking them in glycerine overnight. They may then be examined with a microscope using 100× magnification. Scales may be photographed and later preserved in formaldehyde. No soaking in glycerine is necessary to preserve the scales.

Scales on human hair should be studied under a high-power microscope (400× to 1000× magnification). Place a single hair on a slide and add a drop of water to hold the hair in place. Photomicrographs of human hair may be taken to show the scales, which are arranged around the outermost part of the shaft of the hair. Compare the scales of human hair with scales on hair taken from other mammals such as rabbits, dogs, cats, etc. A variety of mammal hairs can also be obtained from old pieces of fur.

Project 3: Investigating Evolution in the Aquarium

Not all research is done in a laboratory. Libraries, aquaria, museums, and field studies provide fascinating settings for discovery. In this experiment, you can use several of these sources to investigate the principle of evolutionary adaptation among fish.

In all of the great animal groups (the phyla), we find that the organisms evolved in order to adapt to diverse conditions in their environment. Among mammals we see animals that fly, swim, burrow, run, climb, etc. Similarly, among birds there are divers, swimmers, high fliers, and wingless walkers. In all cases the limbs, as well as other parts, have become adapted to the organism's novel means of locomotion.

Here is an interesting aquarium project: study the adaptations of fins among fish. There, and in museums, you will find fish alive or preserved that have explored all the possibilities of life in or close to a water environment. Bring your camera or sketching pad and record fish that live in mud (lungfish), bottom crawlers, flying fish, strong pelagic swimmers, etc. Draw or photograph and study the fin structures; then compare them with the limbs of birds and mammals.

9

A Member of the Master Class of Fishes

THE PERCH

STAND ON THE FOREFRONT ROCKS OF A BREAKWATER AND WATCH the surging schools of minnows contesting the tidal currents, or observe the violent swirling and tearing of the ocean's surface as schools of bluefish slaughter their prey, and you can have no doubt about which animals are dominant in the seas. They are the bony fishes, the *Osteichthyes,* many of which you may have known or eaten most of your life—herring, cod, flounder, tuna, salmon, and others.

The bony fishes are the only fishes that have successfully invaded the fresh waters of the world and established themselves in great populations as prime inhabitants of rivers, lakes, ponds, and streams. The mighty sharks and their close relatives, the rays and skates, have been unable to colonize the freshwater world. The other main group of fishes, the *Crossoptyerigians* (lobe-finned fishes) the lungfishes, represent a small group but a tremendously successful venture from the oceans toward a terrestrial existence.

Most scientists in the field of evolution believe that the first land-living, four-footed creatures, the *Amphibia,* arose from this group some time between 400 million and 450 million years ago, in the Devonian geological period of history. Thus, from this ancient period to the present time the bony fishes have remained the dominant vertebrates in the waters of the world.

While most fishes represent an ancient, vital, and pervasive group of animals, one group, the lungfishes, contributed most directly to the origin of a higher vertebrate group, the *Amphibia.*

An examination of Fig. 100 (p. 156), the Evolutionary Tree of Life, and Taxonomy of the Chordates (p. 157) will reveal the position of the superclass *Pisces* (the Fishes) in relation to the vertebrates in the animal kingdom. The emergence and development of the *Te-*

110

trapoda, four-footed vertebrates, will be seen as associated with the *Dipnoi*, the lungfishes. Before beginning the dissection of the perch (*Perca flavescens*) it would be helpful to review the chapter on the shark (Chapter 8). The anatomy and physiology of the shark is similar to that of the perch.

BONY FISHES AND CARTILAGINOUS FISHES—SOME BASIC DIFFERENCES

The most obvious difference between the cartilaginous fishes, the *Chondrichthyes*, and the bony fishes, the *Osteichthyes*, is the lack of bone in the *Chondrichthyes* group. Both groups have cartilage which appeared earlier than bone tissue in the history of the animal kingdom. Bone tissue emerged in fish from two main sources; replacement of cartilage, "*cartilage bones*"; and *membrane or "dermal bones*," derived from certain scales in the skin. The latter type are generally outside the cartilage bones, closer to the surface, and more readily lifted from the skull. Some of the main differences between the two classes of fishes are listed in Table I.

SETTING THE STAGE FOR DISSECTION

A dissection is more than cutting and finding. It includes an understanding of what is sought and found. It is not enough to locate a gill or some other organ. The arrangement and relationship of the organs and their biochemical interplay, combining anatomy with physiology, breathe life into the organism; and understanding these relationships raises dissection from mere technique to art and science. Therefore, as we proceed with the dissection, some explanations and descriptions of functions will be presented along with directions for the actual dissection.

GETTING THE SPECIMEN READY

Your specimen probably was preserved in formalin. As we have pointed out in previous chapters, formalin can be irritating to the skin, mouth, eyes, and nasal membranes. Therefore, rinse the perch thoroughly with cold water and be sure to have adequate ventilation where you work. The partially dissected specimen should be kept in a plastic bag when the dissection is to be continued at another session.

111

TABLE I. Bony Fishes and Cartilaginous Fishes:
Some Basic Differences

Bony Fish	Cartilaginous Fish
No claspers	Claspers of male (see Fig. 60, p. 97)
No cloaca; separate exit Separate apertures for urine, solid wastes, and sex cells (Lungfish do have a cloaca)	Cloaca present
No rectal gland	Rectal gland present
No true spiral valve (Refer to Fig. 73, p. 117)	Spiral valve present (Refer to Fig. 63, p. 99)
No oronasal grooves	Oronasal grooves present
Air bladder (swim bladder) Rich in O_2; lunglike in some, hydrostatic function in others	No air bladder
Ctenoid, ganoid, or cycloid scales, depending on type of fish	Placoid scales
Endoskeleton ossified (made of bone)	Endoskeleton made of cartilage
Bony operculum covering gills; of dermal origin (formed from scales in skin)	No operculum; gill slits uncovered
Ossification of vertebrae coupled with reduction of notochord	Vertebrae cartilaginous
Pleural ribs formed from dermal bones	No dermal ribs
Skull and jaw bony Bones replace cartilage Some cranial bones, dermal in origin	Skull and jaw cartilaginous No dermal bones
No conus arteriosus at heart, but a thin bulbous arteriosus at base of ventral aorta	Conus arteriosus present
Smaller red blood corpuscles than in shark group (Elasmobranchs)	Red blood corpuscles larger than in bony fishes
Spiracles absent	Spiracles present

FEATURES OF THE PERCH (*PERCA FLAVESCENS*)

The perch is a bony fish belonging to the phylum *Chordata* (see Fig. 100, p. 156). It has a single dorsal hollow nerve chord, evidences of notochordal tissue, and gill slits. It belongs to the *Vertebrata* and has a cranium (skull), gill arches, and vertebral or spinal column. Its notochord has been largely replaced by bony vertebrae, with vestiges of the notochord between vertebrae. It is a member of the Pisces, the fishes, and in that group is an example of the *Osteichthyes*, the bony fishes. Most of the body is covered with scales. An air bladder is present internally (see Table I).

EXTERNAL ANATOMY

Lift the perch from the dissecting pan and hold it in its normal swimming posture. Sight down at the perch and, with the aid of Fig. 71, find and identify the fins. Note that there are paired fins and unpaired median fins. The *median fins* are located in a longitudinal plane that bisects the specimen. The paired fins are the *pectorals* alongside the gill area. The ventral *pelvic fins* are generally more anterior than the pectorals. The median fins are the spiny anterior dorsal and the soft, rayed posterior *dorsal fins*. The *anal fin* is ventral and nearest the *caudal fin*, which is the tail fin. The caudal fin is bilobed and symmetrical. It is known as a *homocercal fin*. The shark has a *heterocercal fin*, that is, it is asymmetrical, with the dorsal part larger than its ventral lobe (see Fig. 59, p. 95).

Remove several overlapping scales. Note that the pigment is in the exposed part of the scale. Examine the scales with a hand lens to see the "growth rings." They can give you a reasonably accurate in-

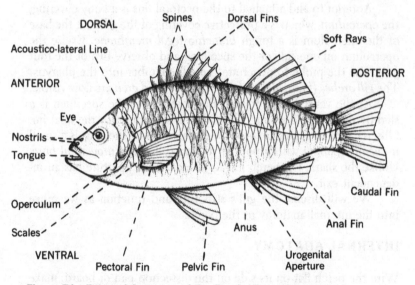

Figure 71. External Anatomy of the Perch: *Perca flavescens*.

113

dication of the age of your specimen. There is little or no ring growth during the fall and winter but rapid growth in spring and summer. Determine how many winters your perch has lived. Perch scales have tiny, toothlike spines at the free ends, called *ctenida*. The scales are known as *ctenoid*.

EXTERNAL FEATURES OF THE HEAD

Place the specimen on the dissecting pan and locate the eyes. Find the *nares* or nostrils anterior to the eyes. The nares lead to an *olfactory* sac. Determine with your probe whether the nares communicate freely with the pharynx.

Open the mouth and find the tongue. Locate the teeth on the dentary bones of the lower jaw. Observe that the teeth point inward, which effectively prevents the escape of prey.

ACOUSTICO-LATERAL LINE

Extending on each side of the body from the head to the tail in a relaxed bow shape is a narrow line called the *acoustico-lateral line*. It communicates with small openings in the scales that lie upon this line. In turn, the terminal aperture in these scales opens to the outside of the body, producing a sensory detection system believed to respond to pressure changes in the water. In some way the lateral line appears to affect body equilibrium in the fish.

Anterior to and attached to the pectoral fins is a bony covering, the *operculum*, which is a protective covering of the gills. At the base of the operculum is a tough *branchiostegal membrane*. Excise the operculum on one side of the specimen and observe two of the four gills. Use the probe to penetrate the gill chamber into the pharynx. The *gill arches* (*branchial arches*) and the *hyoid arch* are now visible.

At the ventral edge toward the posterior of the specimen is a slightly protrusive opening, the *anal aperture*, close to the anal fin. Observe that no scales are near the anus. Just past the anus is a small *urogenital papilla* with a pore at its end called the *urogenital pore*. Unlike the shark, the perch has no cloaca. Each aperture has an independent exit.

We will discuss the gill's structure and function as we move into the internal anatomy of the perch.

INTERNAL ANATOMY

With the perch flat on its side on the dissection pan or board, make an incision with the scalpel from the anal aperture forward, about 1 inch (2.5 cm). Now insert the smooth end of your scissors and cut through the body wall along the ventral line up to the pelvic fin. Grip the cut end with forceps and make a vertical cut at the anal region up to the level of the lateral line. Cut through the body wall, but do not cut deeply or you will destroy internal organs. Make a horizontal incision just below the lateral line and continue the incision until you reach the operculum. With scissors, cut through the body wall from ventral to dorsal at the level of the pectoral fins. Snip off the pectoral fins and continue the cut up to the lateral line. Remove the entire section to observe the *viscera* and the *peritoneal lining*.

EXPOSING THE SEGMENTAL MUSCLES

Cut out a small, shallow flap of skin just below the posterior dorsal fin, down to the lateral line. The flap should be just deep enough to uncover the segmental muscles. Leave the top portion of the flap uncut so that it can drop back into place. The sections exposed are muscles arranged in segments called *myotomes* (see Fig. 72). The myotomes are closely packed and separated from each other by con-

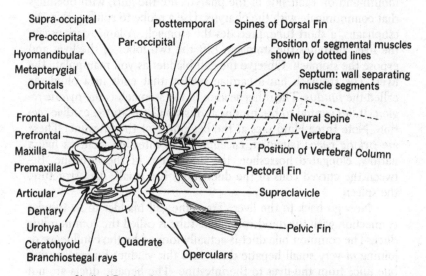

Skull of Perch, showing relationship between skull, vertebral column, position of segmental muscles and skeletal attachment of fins. Most skull bones are shown in diagrammatic form. The Perch skull has about 37 bones, more or less, depending upon interpretation of nomenclature.

Figure 72. The Perch: Skull, Vertebral Column, and Segmental Muscles.

115

nective tissue dividers called *myosepta* (singular, *myoseptum*). Your fish filet dinner is really filet of myotomes.

STRUCTURES IN THE PLEUROPERITONEAL CAVITY

The *pleuroperitoneal cavity* is enclosed by a fine membrane called the *peritoneum*. This membrane will be pierced as you find and examine the *viscera* (internal organs) located within the peritoneal boundary. The membrane directly lining the body wall of the cavity is the *parietal peritoneum*. The internal organs do not float freely in the cavity but are attached to the body wall by membranes that sheathe the organs and connect with the body wall. Such membranes are the *mesenteries* (singular, *mesentery*).

THE DIGESTIVE SYSTEM (FIG. 73)

Beginning at the mouth and proceeding toward the terminus of the digestive system, the anus, find and examine each part of the system. Find the tongue and note its texture. The pharynx is posterior to the mouth and on each side of the pharynx are the gills, with openings that communicate with the pharynx. Use a probe to confirm this. An esophagus, a short tube, precedes the stomach. A large, dark organ, the liver, obscures the stomach. Lift the two lobes of the liver and expose the stomach. Observe the gall bladder as you manipulate the liver. The stomach has a cardiac section that ends in a blind sac called the *fundus*. The other section of the stomach is the *pyloric region*. It lies ventral to the fundus and is smaller than the cardiac section. Note three blind sacs at the beginning of the intestine. These are *pyloric caeca* (singular, *caecum*). The intestine recurves like a narrow, elongated horseshoe, the duodenum of the intestine. Between the curved coils of the duodenum is a small, dark structure, the spleen.

Now go back to the liver. Trace the gall bladder to its tubular connection with the duodenum. The tube is called the *common bile duct*. The common bile duct is actually formed by the confluence or joining of very small hepatic ducts into this viaduct for transferring bile juice from the liver to the intestine. The hepatic ducts are not easily seen. You will be unable to find a pancreas, but there is functioning pancreatic tissue in the perch.

Trace the remainder of the intestine to the anal opening but do not cut away any tissue, or you may destroy parts of the excretory and reproductive organs, as well as blood vessels.

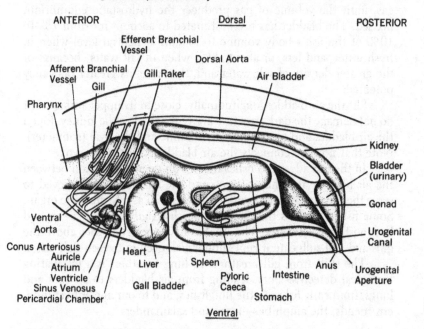

Digestive, Respiratory, Urogenital Systems and Branchial (Gill) Circulation shown. Head, tail and fins omitted, as well as nervous system and endoskeleton. Branchial arches circulation exaggerated to promote clarity.

ANTERIOR   Dorsal   POSTERIOR

Efferent Branchial Vessel
Dorsal Aorta
Afferent Branchial Vessel   Gill Raker
Gill
Air Bladder
Pharynx

Kidney

Bladder (urinary)

Gonad

Ventral Aorta

Urogenital Canal

Conus Arteriosus
Auricle
Atrium
Ventricle
Sinus Venosus
Pericardial Chamber
Heart
Liver
Spleen
Gall Bladder
Pyloric Caeca
Stomach
Intestine
Anus
Urogenital Aperture

Ventral

Figure 73. Internal Organs of Bony Fish: The Perch.

TO DIVE, RISE, OR STAY LEVEL

Just above the intestine is a large, conspicuous, whitish sac that looks like a plastic, air-filled bag (see Fig. 73). this is the *air bladder* or *swim bladder*, and it is indeed gas-filled. A fish living in water has to adjust to its degree of buoyancy in the water. A fish without an air bladder will sink because it is denser than the water around it. To swim at different levels, a fish has to decrease its density proportionately, to enable the water's buoyancy to keep it from sinking. The air bladder serves this purpose. Bottom fish lack an air bladder.

The air bladder is capable of secreting oxygen by means of a special group of blood vessels called the red gland or *rete mirabile*, situated at the anterior part of the bladder. The more oxygen secreted into the bag, the more buoyant the fish becomes, and it will rise to a higher level, aided of course by its muscles and fins. When it secretes enough oxygen to prevent further rising or dropping, it will maintain its balance on an even keel. To change its level it must add

117

more gas or lose some of it. The latter is accomplished by the *oval*, a special oxygen-absorbing unit with a sphincter, located at the posterior of the air bladder. This unit absorbs oxygen, thus discharging the gas, until the volume of gas produces the hydrostatic equilibrium needed. The bladder has been estimated to account for from 7% to 10% of the fish's body volume to remain at neutral level when in fresh water, and less, or about 5%, when in salt water, because of the greater density of salt water and, therefore, its greater buoyancy potential.

Slit the air bladder longitudinally, close to its upper surface, but do not damage the dark structure above it, which is the kidney. Open the air bladder and locate the rete mirabile and the oval (sphincter). Note that no tube connects the air bladder with the pharynx.

In the less advanced fishes there is an open connection between the air bladder and the pharynx, and the air bladder is believed to serve these fishes as a respiratory organ as well as a hydrostatic organ. Some fishes, such as the climbing perch (*Anabas*), breathe air and live mostly out of water. The Anabas have special air sacs above the gills and will suffocate if kept submerged, even in oxygen-rich water.

The shift from gill breathers to lung breathers is a fascinating biological detective story, leading from air bladders to air sacs and lungs, from early fishes to the lungfishes, and to our ancient yet modern friends, the amphibia—frogs and salamanders.

THE EXCRETORY SYSTEM—MAINTAINING THE SALT AND WATER BALANCE (FIG. 73)

Marine fishes live in a saltwater environment. The concentration of salts, especially sodium chloride, is greater in the water than in the fishes' blood and tissues. Therefore the marine fishes have a problem: how to prevent loss of water from blood and tissues to their aquatic medium and how to prevent gaining excess salts in their blood and tissues. This is partially solved by special chloride-secreting cells in the gills that dispose of excess salts. These cells make it possible for marine fishes to drink some salt water. Additionally, mucus secretion around the fishes' skin tends to prevent absorption and loss of body water.

Freshwater fishes have the opposite problem. Their blood and body tissues have a higher salt concentration than the water of their environment. Their problem is to conserve salt and prevent excessive retention of water. This is partially solved by the kidneys, which have special blood vessels and tubules arranged to form *glomeruli*. The glomeruli can reabsorb salts into the blood and remove excess water from the blood through the urinary duct, and out of the body through the urogenital pore. The glomeruli in the kidneys of fresh-water fishes are far more numerous than in the kidneys of marine fishes. Thus, the marine fishes tend to lose less water from the kidneys than do the freshwater fishes.

Actually, the gills serve two major purposes in fishes: respiration and excretion. They aid in controlling the salt-to-water balance, excreting urea and ammonia, which are protein breakdown products; and they aid in absorbing oxygen. We will examine the respiratory function of the gills after completing our discussion of the circulatory system, which is critically involved with the gills and the kidneys.

DISSECTION PROCEDURES FOR LOCATING THE KIDNEYS

Between the parietal peritoneum and the dorsal surface of the air bladder lie the brownish kidneys, one on each side of the body (see Fig. 73). Use a probe to trace the narrow kidneys to their emerging, posterior urinary tubes. Note that the two tubes fuse to form a larger sac ventrally, the urinary bladder. Trace the course of the duct leading from the bladder to the urogenital opening, from which urine passes out of the body.

THE CIRCULATORY SYSTEM—PICKUP AND DELIVERY

Review the circulatory system of the shark and study Figs. 65, 66, and 67 in Chapter 8. The perch's circulatory system is very similar to the shark's circulatory system. The circulatory system in both the perch and the shark serves the same function: pickup and delivery to all parts of the body of transportable oxygen and other needed chemicals, and delivery of tissue wastes to the body terminals for their disposal.

PARTIAL DISSECTION OF THE CIRCULATORY SYSTEM

We will concentrate on exposing the heart and its directly associated blood vessels: blood vessels bringing blood into and out of the heart. In addition, we will examine the branchial or gill circulation, where oxygenation of blood and removal of carbon dioxide, urea, and ammonia from the blood take place.

119

If you wish to trace the main blood routes of the perch, simply follow the directions supplied in the chapter on the dogfish shark, locating the tributaries of the dorsal aorta, and the return of the blood through the veins of the renal portal circuit, the lateral abdominals, and the cardinal veins.

OPENING THE PERICARDIAL SAC

The original ventral incision starting from the anal aperture ended at the pelvic fin area. Now continue that ventral incision, cutting through the body wall below the gills and past the forward end of the gills. Carefully cut dorsally at each end of the new incision, about up to the lowest level of the pharynx (see Figs. 73 and 74). The heart is encased in a sac bounded by a tough membrane called the *pericardium*. Note the vessels entering and leaving the sac and slit the membrane to expose the heart and the vessels connected with it.

The heart has two main chambers, a thickly muscled *ventricle* and a thin-walled *auricle*. Connected with and opening into the auricle is a thin sac, the *sinus venosus*. Blood vessels returning blood to the heart open into the sinus venosus. These are the *common cardinal veins*, also known as the *Ducts of Cuvier*, and the *hepatic vein* from the liver. The direction of flow of the blood is from the common cardinal veins and hepatics to the sinus venosus and from the

Source: *Life of Vertebrates*, J.Z. Young, Oxford U. Press, 1981

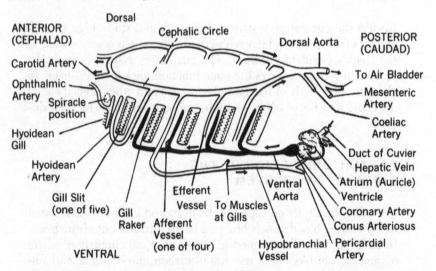

Figure 74. Branchial Circulation of Bony Fish.

120

sinus venosus into the main receiving chamber, the auricle. The blood is then moved into the ventricle as the auricle contracts, though not powerfully, since the walls are thinly muscled. The ventricle provides the power thrust to the blood, driving it into a short, thickened *bulbus arteriosus*, actually the beginning of the ventral aorta. The bulbus provides an extra fillip of thrust to the movement of the blood.

BRANCHIAL CIRCULATION—HEART-TO-GILLS PASSAGE

Use scalpel, scissors, and forceps to clear tissue away from the blood vessels. Run the edge of the probe along the exposed blood vessels to help trace their course. Starting with the large ventral aorta (see Fig. 74), remove tissue adhering to the branches of the ventral aorta and expose the four *afferent branchial arteries* that deliver blood to the gills. Carefully locate and identify the *hypobranchial vessel* that lies directly below the ventral aorta and rises to its origin at the *dorsal cephalic loop*. This vessel is interesting because it supplies oxygenated blood to the heart muscles and to the muscles of the gills.

Alongside the afferent branchial arteries find the *efferent branchial arteries*, which carry oxygenated blood. The blood reaching the heart carries deoxygenated blood from the tissues of the body. When blood passes from the heart into the ventral aorta it still has deoxygenated blood. The blood becomes oxygenated only after the capillaries linking the afferent with the efferent branchial arteries in the gills absorb oxygen from the water as it passes through the gills. Here the stage is set for the vitally needed oxygen to be delivered to all parts of the body. The efferent branchial arteries move blood into the dorsal aortic or *cephalic circle*, from which arterial branches conduct the oxygenated blood forward to the head area and backward to the torso and caudal area. Of course, to complete the cycle, the blood returns by means of veins to the heart, after yielding its cargo of oxygen to the cells and organs of the body, and picking up metabolic or cellular wastes for ultimate disposal. Thus the circle of blood is joined.

THE GILLS AND RESPIRATION—UNDERWATER BREATHING

Each side of the pharynx has four gills (see Fig. 73). Remove one of the gills and examine it with a hand lens. Observe that on each gill there are double rows of feathery-looking structures called *gill fila-*

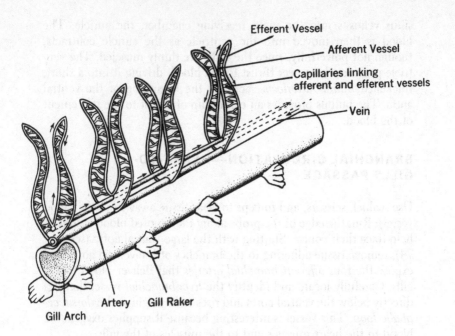

Efferent Vessel

Afferent Vessel

Capillaries linking
afferent and efferent vessels

Vein

Artery   Gill Raker

Gill Arch

Expanded segment of a gill showing structure of gill. Filaments
shown spread apart for clarity. They are densely clustered in the
actual gill. Direction of blood flow shown by arrows. Inner vessels
are *afferent* (blood flows from ventral aorta toward gills). Outer
vessels are *efferent* (blood flows from capillaries to efferent vessels and away from gills). The gill arch is a bony support of the
gills. Gill rakers act much like the teeth of a garden rake to stop
and impede passage of large particles through gill slits from pharynx. The gill rakers are on the inside of the gill arch, the pharynx
side.

Figure 75. Expanded Segment of a Gill.

ments (see Fig. 75). Snip off one filament and examine it with a microscope, using the lowest-power objective. Locate the fine blood
vessels—the capillaries—that connect the afferent branchial artery
on one side of the filament to the efferent branchial artery on the opposite side. Observe the epithelial cells that line the entire filament.
The bridge of capillaries is the site of the absorption of oxygen
needed for cellular oxidation. It is also the locus for the release of
carbon dioxide, a waste product of cellular oxidation.

Note that the gill filaments rest upon a tough but resilient cartilaginous rod, the *gill arch.* Find the *gill rakers* attached to the inner
surface of the gill arch. The rakers prevent food or debris that enter
the pharynx from getting through to the delicate gill filaments.

122

HOW THE FISH BREATHES UNDER WATER—
THE MECHANICS OF RESPIRATION

To obtain the life-supporting oxygen from the ocean, the fish must create a pressure differential between the outside water and the water inside its mouth. This requires the following conditions:

1. A decrease in the internal pressure within the mouth and pharynx, causing a surge of water inward.
2. A closing of all internal exits and an enlargement of the pharyngial cavity, thereby reducing the internal water pressure.
3. Prevention of a backflow of water from the pharynx toward the mouth.
4. Muscular pressure and synchronous timing of the opening and closing of valves that direct the flow of water.
5. Forceful pumping of water toward the gill exits, with no reverse flow.

These are some of the mechanical problems the fish has to solve in order to get the oxygen it requires to sustain life. How are these conditions achieved?

THE WATER INTAKE PHASE

With the fish's mouth open and no internal adjustments, water will be in the mouth, but no water will flow inward under pressure, since there is a neutral differential of pressure between the external and internal environment. Now the dorsal and ventral oral flaps in the buccal (cheek) area open. The operculum closes, moving inward against the gill slits. The gill arches are pressed laterally, enlarging the cavity of the pharynx and thereby decreasing the internal pressure. Meanwhile, the branchiostegal membranes below the operculum prevent inflow of water from outside the gills. The combined effect of these events produces the correct water pressure gradient to cause a surge of water inward. The entire process of intake and outflow of water is influenced by neural control centered principally in the medulla oblongata of the brain.

THE WATER OUTFLOW PHASE

With the mouth closed the gill arches retract, increasing pressure in the pharynx. The operculum opens, enabling flow of water to and through the gills. The oral flaps close off the pharyngeal chamber and muscles aid in driving the water in a stream across the gills' filaments, like water flowing over a filter bed. The capillaries in the filaments are thin-walled, allowing for osmotic exchanges and absorption of gases to take place. The hemoglobin in the red blood corpuscles circulating within the capillaries absorbs the oxygen from the water and combines with it to form the temporary oxygen carrier, the oxyhemoglobin molecules. The intake–outflow cycle is now completed, and will be repeated for the life of the fish.

THE NERVOUS SYSTEM—THE CHAIN OF COMMAND

The role of the nervous system is to keep the fish in constant touch with its external and internal environments and to trigger actions that will help the fish to survive. How it succeeds, or sometimes fails, is an enormously complex story whose leading characters are known. But the finer details of the plot have yet to be learned.

The nervous system of the bony fish is like that of the shark (study Figs. 76a and 76b). It consists of two major divisions:

1. The *central nervous system*, consisting of a *brain* and ten pairs of *cranial nerves*, and a *spinal cord*, with segmentally paired *spinal nerves*;

2. An *autonomic system* with two divisions—*a parasympathetic system* and a *sympathetic system*—made up of *sympathetic ganglia* and *sympathetic nerves* connected with muscles and glands at one end of the nerves and with the brain and spinal cord at the other end.

Sharks have no sympathetic system in the head. Higher forms of animals have a highly developed branch of the autonomic nervous system, the parasympathetic system. But our knowledge about the parasympathetic system of the perch and other bony fishes is limited.

EXPOSING THE BRAIN

Before cutting through the cranium to find the brain, study Fig. 72. This will enable you to identify some of the major bones in the head. Remove some of the bones' overlying tissue with a sharp scalpel, using shallow, flat cuts. Dorsally, find and identify the occipital, preoccipital, supraoccipital, and paroccipital bones. Also locate the prefrontal, frontal, and orbital bones. Looking laterally at the fish's

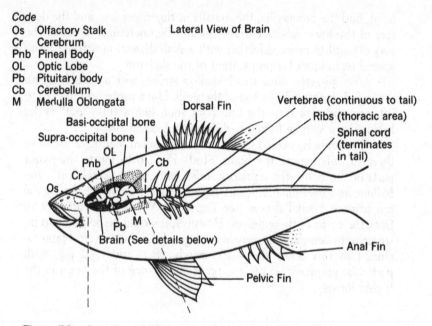

Code
Os Olfactory Stalk
Cr Cerebrum
Pnb Pineal Body
OL Optic Lobe
Pb Pituitary body
Cb Cerebellum
M   Medulla Oblongata

Lateral View of Brain

Basi-occipital bone
Supra-occipital bone
OL
Pnb
Cr
Os

Dorsal Fin

Cb

Vertebrae (continuous to tail)
Ribs (thoracic area)
Spinal cord (terminates in tail)

Pb
M
Brain (See details below)

Anal Fin

Pelvic Fin

Figure 76a. Location of Brain and Spinal Cord in Bony Fish.

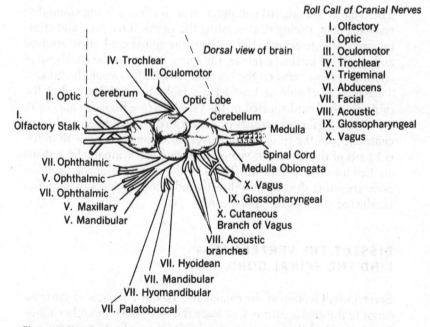

Roll Call of Cranial Nerves
I. Olfactory
II. Optic
III. Oculomotor
IV. Trochlear
V. Trigeminal
VI. Abducens
VII. Facial
VIII. Acoustic
IX. Glossopharyngeal
X. Vagus

Dorsal view of brain

IV. Trochlear
III. Oculomotor
II. Optic
Cerebrum
I.
Olfactory Stalk

Optic Lobe
Cerebellum
Medulla
Spinal Cord
Medulla Oblongata

VII. Ophthalmic
V. Ophthalmic
VII. Ophthalmic
V. Maxillary
V. Mandibular

X. Vagus
IX. Glossopharyngeal
X. Cutaneous Branch of Vagus
VIII. Acoustic branches
VII. Hyoidean
VII. Mandibular
VII. Hyomandibular
VII. Palatobuccal

Figure 76b. Projection of Brain from Fig. 76a to Show Major Parts and Location of Cranial Nerves.

head, find the premaxilla, the maxilla of the upper jaw, and the dentary of the lower jaw. Ventrally, locate the operculars. It would be very difficult to proceed further with a skull dissection without using special techniques for preparation of the skeleton.

Very gingerly, using small, shallow strokes with a scalpel, cut an oval window through the top of the skull. Use a probe and forceps to lift, pry, and pick away the thin bony roof. Below this "roof" is the brain. Use a syringe to gently flush away loose tissue.

The brain is covered with membranes called *meninges*. Remove these to fully expose the brain. Study Fig. 76 to locate the major parts of the brain: the *olfactory stalks, optic lobes, cerebrum, cerebellum,* and *medulla oblongata.* Arising from these structures are the ten pairs of cranial nerves (see Fig. 76b). We will not attempt to trace the paths of these nerves. However, it would be helpful and instructive to compare the nervous system of the several vertebrates included in this book: the shark, perch, frog, and fetal pig, with particular emphasis on the emergent dominance of the brain in the higher forms.

THE BACKBONE OF COMMUNICATION

The brain is the original computer. It is in effect a living computer: receiving data, storing it, associating the signals it receives, and creating integrated responses. But without the spinal cord safely encased and protected by the vertebrae, the spinal column—the backbone of communication between the head and the body—would be broken. Therefore we should at least have a look at a *vertebra* and at the cable of spinal cord encased by the vertebrae (see Figs. 77a and 77b).

The spinal cord consists of a great many nerve fibers and nerve centers. From the spinal cord, nerves make connections with different parts of the body. The sympathetic and parasympathetic systems are tied into the spinal cord. Thus, all the muscles and glands of the body are either directly or indirectly influenced to react by impulses conducted through the spinal cord.

DISSECT THE VERTEBRAE—
FIND THE SPINAL CORD

Sever a small section of the caudal area (about 2 inches—5 cm) posterior to the anal aperture. Cut away the myotomes and other adhering tissue until the vertebrae are visible. If you find a spinal nerve, whitish in color, try to trace it, undamaged if possible, to its connec-

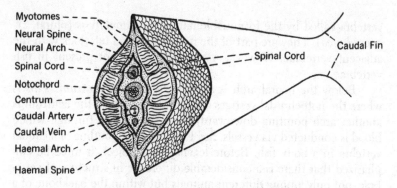

Figure 77a. Tail Section of the Perch Showing Caudal Vertebra.

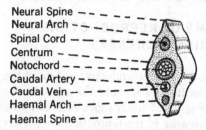

Figure 77b. Caudal Vertebra of a Perch.

tion with a vertebra (see Fig. 77). Continue removing tissue with forceps from the exposed vertebrae on one side and repeat the process on the other side of the severed section. Aim at freeing a series of three adjacent vertebrae to see how they articulate (join) with each other. Separate one from the series and examine the slender tissue pad between them. This tissue represents the vestiges of the *notochord* found in all vertebrates in the embryonic state. Intervertebral tissue is also found between the vertebrae of humans. It is notochordal, and each tissue pad is known as a *disc*. The disc is important in preventing physical damage to the bones during vigorous movements like jumping, weight lifting, and walking. It acts to reduce friction between the vertebrae, preventing chipping of the edges of the spinal bones as they glide against each other during bodily movements.

Use your hand lens to see the cut ends of the spinal cord on the front end of the vertebra and the other at the opposite end. The spinal cord passes through the *neural canal*, which is in the *neural arch*, the dorsal part of the vertebra. Find the neural spine on top of the neural arch.

Each vertebra forms a joint with its neighbor. In higher animal forms the joints are strengthened by small bony extensions of the vertebra called by the formidable name of *zygapophysis* (plural, *zygapophyses*). They are part of the neural arch and interlock with the adjacent vertebrae. However, zygapophyses are not present in fish vertebrae.

Below the neural arch is a round, biconcave *centrum*. This is where the notochordal vestiges are found. Ventral to the centrum is another arch pointing downward, the *haemal arch,* through which blood is conducted via vessels. See Fig. 77b for an outline of a caudal vertebra in a bony fish. Before leaving this subject, it must be emphasized that there is a considerable difference in structure of vertebrae, not only among different animals but within the backbone of a single animal. Cervical vertebrae differ from thoracic, lumbar, and caudal vertebrae, and each has its own structural differentiations.

THE ANATOMY OF REPRODUCTION— HOW THE TRIBE INCREASES

Dissection to inspect the reproductive system of the perch is unnecessary. Simply locate and identify the reproductive organs of your specimen. (see Fig. 73). Merely move aside any structures that interfere with clear viewing of the reproductive structures.

No external structures differentiate the male from the female. Internally, the female has two *ovaries,* which are united as one and appear bilobed. They are located in the abdominal cavity (pleuroperitoneal cavity) attached to the dorsal wall by mesenteries, and extend posteriorly and ventrally. Eggs (*roe*) are shed during the mating season and exit the body through the urogenital pore near the anal opening (see Figs. 71 and 73).

The male *gonads* (*testes*) are small and narrow in the nonreproductive seasons and enlarge during the spawning season. They are located in the abdominal cavity above the intestine and are moored to the dorsal wall of the cavity anteriorly, by mesenteries. Male *gametes—sperm cells—*are produced in the testes and released with a *seminal fluid* as *milt* over the eggs during the mating process. The sperm cells leave the testes through *vas deferens ducts* and are discharged, through the urogenital papilla, into the water. There fertilization takes place and, warmed by the sun, the *zygotes* (fertilized eggs) rapidly begin the marvelous sequence of vital divisions and changes that appear to mirror the biological phases of their ancestry.

PRELUDE TO THE TETRAPODA

After dissecting two representatives of the superclass Pisces—the cartilaginous shark and the bony perch—we will continue to climb the evolutionary tree of life (see Fig. 100, pg. 156). While some fish ventured successfully into an environment strange and hostile to water dwellers, their success as pseudoterrestrial organisms was limited. Except for limbless snakes, which indeed had ancestors with legs, no vertebrate could conquer the challenge of terrestrial life without lifting its body from the earth with supporting appendages, the legs. Having descended from an ancestry of gill breathers, even overcoming the problem of breathing air was insufficient to promote a victorious invasion and colonization of terra firma. Those animals whose genetic makeup changed to produce forelegs, hind legs, or wings could become the hunters or the surviving refugees of the hunters.

The story of these fascinating developments is part of a history of life which involves a great deal more than dissection. But we can begin to appreciate the differences between the superclass Pisces and the superclass Tetrapoda, as we move on in the next chapter to the dissection of the frog, a very successful and ancient tetrapod.

Project 1: The Cardiac Concern

In recent years people have become more and more interested and actively involved in the prevention of heart attacks and circulatory problems. Our attention and admiration has been heightened by remarkable developments in heart surgery, the use of a mechanical heart, heart transplants, and new chemicals to treat cardiac and circulatory conditions. The advances have been made possible by experimentation. Generally, initial experiments are conducted on different animals. The specific results of such experiments may not entirely parallel or duplicate the effects on humans. Nevertheless, because the biochemistry and much of the tissue structure have elements common to humans, the results of animal experimentation are often reasonably indicative of what to expect in humans. We owe a great deal to the scientists and the subjects used in the crusade against disease.

A challenging question is, "How far back on the evolutionary road will such testing bear a real biological relevance to humans?" In this project we will work backward. Start with chemicals whose action on the heart or other phases of human circulation is somewhat known, and investigate the effects on lower vertebrates to discover reaction relationships.

Reverse Animal Experimentation. Here is a relatively uncomplicated method of exploring the effects on fish of chemicals known to affect human heart and circulation. We will suggest several chemicals and a method for testing them. But bear in mind that this project is open-ended. You can substitute a tadpole for a fish or any

129

other organism, with suitable modifications of technique and chemicals used. Remember also to consider that the mass of the tested organism will undoubtedly be far less than that of a human, and therefore the dosage should be proportional. Also, be very careful to *work under supervision,* since some chemicals, *indeed most chemicals, can be dangerous or even lethal if used improperly.*

Observing the Effects of Several Household Medicines on Fish. Assemble the following:

Equipment	Suggested Medicines (substitute others if you wish)
Microscope	Aspirin
Glass square (about 3″—7.5 cm—square)	
Cotton gauze	Tylenol
2 large jars (preferably battery jars)	Nasal decongestants (your choice)
	Eye drops (your choice)
Rubber bands (about 4″—10 cm—long)	*Under supervision only:*
Graduated pipette or hypodermic syringe, 2 cc	Nitroglycerin pills 1/150 mg
	Inderal, 10 mg per pill

We will set forth the technique for preparing the fish to observe its blood flow under the microscope and to suggest how to apply the chemicals you will use. The rest is up to you.

How to Prepare the Subject. Keep the fish selected (guppy, small goldfish, or any other small specimen) in a battery jar, with water at room temperature, after the chemicals have been prepared for testing. When ready for testing, wrap the body of the fish with water-saturated cotton, leaving the mouth opening and the tail exposed. Place the wrapped fish on the glass square and cover the saturated cotton with a swath of cotton gauze cut to fit the size of the specimen. Be careful not to cover the tail or head. Slide a rubber band, *not tightly but firmly,* at each end of the gauze, to hold the fish in place and to prevent its flopping about.

Examine the tail under low power on the microscope to find the blood circulating in the caudal vessels. Observe the character of the blood flow; smooth, continuous, or in spurts. Try to establish a rate of flow between two points by fixing on corpuscles moving from the initial point to the farthest point in the field of vision. Perform this several times at daily intervals, allowing the subject to recover com-

130

pletely. Keep detailed and careful records. Replace the fish in a tank or battery jar, after removing the gauze and cotton. The purpose of repeating this procedure is to help you to achieve some measure of standardization of the routines and observations as a control for comparison with the testing effects.

Preparing the Chemicals for Testing

Step 1. Grind the tablets to a fine powder in a mortar with a pestle. Then mix with water until most of the powder is dissolved or has formed a fine, faintly cloudy colloidal mixture. (A colloid is a mixture of a solute [the substance being dispersed] in a solvent [the dispersing medium]. The solute is not dissolved, but its particles are so small that they remain evenly dispersed in the solvent and do not settle to the bottom.) This will approximate a saturated condition.

Step 2. Draw up a small amount in a medicine dropper. One drop is generally 1 cc. If available, use a graduated pipette or a hypodermic syringe to better control the volume of solution you will administer to the subject.

Step 3. Liquids—nasal decongestants or eye drops—should be drawn up by medicine dropper, pipette, or hypodermic and applied without dilution, minimally to begin with, 1 ml or one drop at a time for each test condition, until you can view some alteration in the fish's circulation.

Some chemicals, such as nasal decongestants and eye drops, act by causing vasoconstriction of the blood vessels in humans. Others, such as nitroglycerin, act oppositely, causing vasodilation and resulting in a greater volume of blood flow through the blood vessels.

Application of the Chemicals to the Specimen. Be sure to keep the subject thoroughly moist with water. Set the fish in position for viewing the circulation of blood. Observe the normal pattern. Now add one drop of the test chemical at the gill area and another at the mouth. Remember, the operculum of your subject may be closed at the moment of application. Try to time the application of the test chemical with the movement of the operculum, so that the gills may be exposed for rapid absorption of the chemical into the blood. Record your observations and resourcefully refine your techniques to produce as valid results as you can manage.

Project 2: Disappearing Through Camouflage

Obtain several small, live flounder (about 2 inches—5 cm—long). Keep them in a marine aquarium. (See Project 1 in Chapter 7, "The Starfish.") Use a second, smaller, all-glass tank, 2-gallon size, for the following experiment.

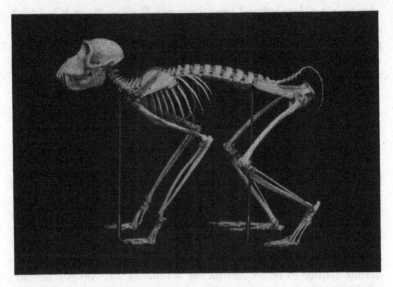

Figure 78. Rhesus Monkey Skeleton. (Photograph by Carolina Biological Supply Company)

Figure 79. Opossum Skeleton. (Photograph by Carolina Biological Supply Company)

132

Place colored paper around the bottom and three sides of the tank. Transfer the flounder to the small tank for 2 to 4 hours. Observe the color changes that take place. This may help answer the question: "Why is a flounder so hard to see in its natural surroundings?" Change the color patterns every 2 or 3 days. Record the effect of these changes on the skin color of the flounder. Can you tell what the advantages of these color changes are to the flounder? Some fascinating work has been pioneered in this field.

Project 3: How to Tell a Tale from Head to Tail

Obtain vertebrae (individual sections of the spinal column) from members of the main groups of vertebrates: fish, amphibia, reptiles, birds, and mammals (see Figs. 78 and 79). A visit to a museum will give you a chance to photograph vertebrae of prehistoric animals such as dinosaurs.

Arrange the vertebrae and the photographs to show group relationships from the simplest group of vertebrates, such as the fish, to the most complex groups, such as birds and mammals.

Visit the library to get books on comparative anatomy and learn the differences in vertebrae structure. Now select any vertebra and build a true tale around it. See how good a science detective you are by answering the following questions: How large was the animal to whom the vertebra belonged? Was it a quadruped? A bird? A bat?

We will next dissect a frog, a vertebrate tetrapod.

10

THE FABULOUS FROG

MAN OWES A GREAT DEBT TO THE FROG. FOR GENERATIONS THE frog has contributed to our knowledge of anatomy, physiology, embryology, parasitology, and many other of the important "ologies." This lowly vertebrate, almost comical in appearance, has been an old standby among scientists, in the school classroom, and in the professional laboratory, and will continue to be a well from which to draw knowledge of the science of life. It is an animal that survives well in captivity and is easily handled.

The frog leads a double life. As a tadpole, before maturity, it is an aquanaut, a water dweller breathing by gills; as an adult it is a terranaut, a lung-breathing land dweller. However, because its eggs are laid in water and because there are no protective scales on its skin, the adult frog must live in moist surroundings near a source of water. Skin divers have learned from the frog, with its webbed toes, how to propel themselves rapidly in water. In fact, skin divers are sometimes called frogmen.

The class of vertebrates, *Amphibia*, of which the frog is a member, includes salamanders, toads, and wormlike creatures called caecilians. They are an ancient and venerable group; one might say they are almost living fossils. Once a great and dominant group, they are now the smallest group of vertebrates. Their legacy to the progress of evolution was to serve as ancestor of the reptiles which, in turn, gave rise to birds and mammals and finally to man, a representative of the mammals. Now their evolutionary job appears to be done. Even today there is evidence of their ancient greatness. Japan has a salamander that grows to five feet in length and weighs up to 100 pounds (about 45 kg). At the other end of the scale is a Cuban frog, all of 3/8 inch (about 1 cm) long.

Certain of the amphibia are of great economic value to humans. For instance, the giant South American toad can eat, in one month, up to 3000 or more insects harmful to man. For this reason the South American giant toad has been exported to the Hawaiian Is-

134

lands to combat the grubs that destroy sugar cane. Another example of the frog's value is evidenced by the practice of stocking lakes with tadpoles in mosquito control programs, because they eat mosquito larvae or wrigglers.

In many respects the anatomy of the frog is similar to human anatomy. A study of the frog's anatomy gives us a good idea of the main features of human anatomy. It also reveals basic evolutionary differences between the structure of a cold-blooded, fairly primitive vertebrate and the structure of an advanced, warm-blooded vertebrate like man. It also highlights the difference between a true amphibian with a three-chambered heart and a true lung-breathing terranaut, like man, with his four-chambered heart.

LET'S PREPARE FOR THE DISSECTION

Obtain a double-injected male and egg-bearing female American Bullfrog (see Fig. 80), *Rana catesbiana,* the largest American frog, or the leopard frog, *Rana pipiens,* smaller and less costly. Do not get the pickerel frog (which looks like the leopard frog) because its skin produces a secretion that causes allergic reactions in some people, resulting in inflammation of the skin on the hands.

You will require the basic dissection equipment used for the earthworm, plus a pen flashlight or pen flashlight magnifier, a wooden matchstick, a pair of strong 3½-inch (about 9-cm) scissors with one blunt tip and one sharp tip, and a bristle from a nylon hairbrush.

Figure 80. American Bullfrog *(Rana catesbiana).* (Photograph by Carolina Biological Supply Company)

EXPLORING THE EXTERNAL ANATOMY
OF THE FROG

Place the frog dorsal side up in the dissecting pan. Stroke the skin with your finger and feel its smoothness. Examine the skin with your hand lens to note the absence of scales. The frog breathes through its moist skin, diffusing oxygen into the tiny blood vessels in the skin. It also breathes by means of lungs (see Fig. 94 later in this chapter).

To familiarize yourself with the frog, use Fig. 81 as a guide and identify the labeled parts of the diagram on your specimen. Then turn the specimen ventral side up and notice the difference in coloration on its ventral side. These color differences and color patterns help the frog to blend into the background of his natural habitat on land and in water. This protective camouflage has prevented many a frog from becoming a tasty tidbit for a snake.

Let's look into that awesome cavern, the frog's mouth. Fig. 82 (a and b) shows how to prepare the specimen for the study of the

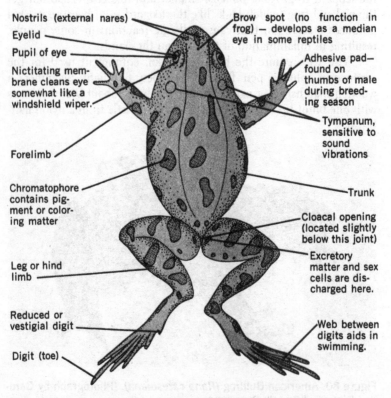

Figure 81. External Anatomy of the Male Frog—Dorsal View.

136

structures in the mouth cavity. Insert a fine, firm strand from a nylon hairbrush into one of the Eustachian tubes and watch the *tympanum* or eardrum on the dorsal side of the frog. You will see the tip of the strand pressing upward against the tympanum. This primitive hearing system allows air pressure to be equalized in the head of the frog. It acts as an aerostat. Your male specimen will have two vocal sac

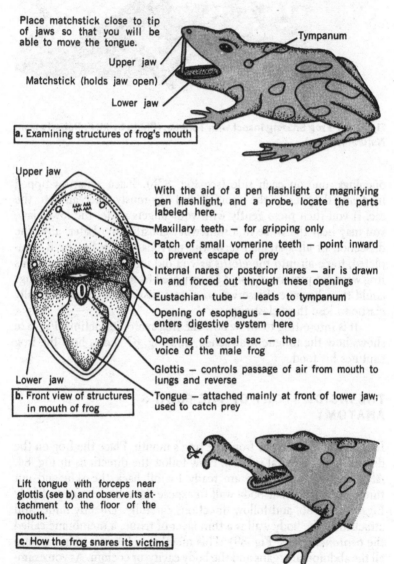

Place matchstick close to tip of jaws so that you will be able to move the tongue.

Tympanum

Upper jaw

Matchstick (holds jaw open)

Lower jaw

a. Examining structures of frog's mouth

Upper jaw

With the aid of a pen flashlight or magnifying pen flashlight, and a probe, locate the parts labeled here.

Maxillary teeth — for gripping only

Patch of small vomerine teeth — point inward to prevent escape of prey

Internal nares or posterior nares — air is drawn in and forced out through these openings

Eustachian tube — leads to tympanum

Opening of esophagus — food enters digestive system here

Opening of vocal sac — the voice of the male frog

Glottis — controls passage of air from mouth to lungs and reverse

Lower jaw

b. Front view of structures in mouth of frog

Tongue — attached mainly at front of lower jaw; used to catch prey

Lift tongue with forceps near glottis (see b) and observe its attachment to the floor of the mouth.

c. How the frog snares its victims

Figure 82. The Mouth of the Frog.

137

Figure 83. Frog Snaring Insect with Tongue. (Photo by M. F. Roberts from *Natural History*)

openings, one on each side (see Fig. 82b). Place a narrow-tipped medicine dropper at the opening and vigorously force air into the sac. If you then press gently with your fingers under the lower jaw, you may hear the voice in a somewhat ghostly way. Later, after the dissection of all systems except the nervous system has been completed, force air into the *glottis* as you did into the vocal sac. If this frog were not preserved, you'd be surprised to see how large the lungs could then become. Preservation causes tissues which are normally elastic to lose their elasticity.

It is interesting to know how the frog snares its victims. Fig. 82c shows how the tongue is attached, and Fig. 83 shows how the frog captures his food.

THE INSIDE STORY OF THE FROG'S ANATOMY

Remove the matchstick from the frog's mouth. Place the frog on the dissecting pan ventral side up. Now follow the directions in Fig. 84. After this is finished we are ready to roll back the skin and cut through the muscular body wall to expose the internal organs. Study Figs. 85 and 86, and follow directions carefully. Directly under and attached to the body wall is a thin layer of tissue, a membrane called the *peritoneum* (see Fig. 89). This membrane forms a lining around all the abdominal organs and the body cavity or coelom. As you examine the internal organs you will see membranes that hold organs to-

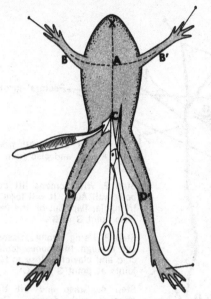

Step 1. Pin frog to dissecting pan ventral side up as shown.

Step 2. Lift skin with forceps at point C and make shallow cut through lifted skin with a razor or scalpel.

Step 3. Insert rounded edge of scissors into incision and cut skin along dotted line to point X.

Step 4. Make lateral cuts from point A to B and B¹. Then cut down from point C to D and D¹.

Figure 84. First Steps in Opening Body Cavity of the Frog.

Step 1. Using forceps lift skin as shown at point E and with scalpel cut skin free from F to D¹ and similarly from B to D. Pull skin back firmly with forceps as tip of scalpel cuts it free. The unattached spaces under skin are lymph spaces (important in circulation).

Step 2. Free the skin on both sides so as to expose the body wall. Roll all skin flaps back and pin each side to pan.

Step 3. Cut a small opening through muscular body wall at point C with scalpel or razor. Hold wall with forceps as you cut.

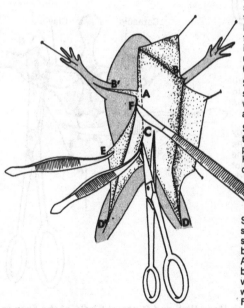

Step 4. Insert rounded tip of scissors into incision, holding scissors almost parallel to body. Continue cutting to point A where sternum (breastbone) begins. **Caution:** cuts must be very shallow or internal organs will be damaged. Now turn to Fig. 86.

Figure 85. Opening Body Wall of the Frog.

139

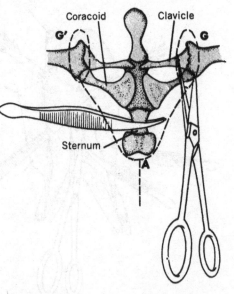

Pectoral girdle

Step 1. Cut body wall from point A to G and also from A to G[1].

Step 2. With forceps lift cut body wall AGG[1]. It will form a triangular flap. Cut off the flap from point G to G[1].

Step 3. Using strong scissors cut through the bones (coracoid and clavicle) close to the joints at point G and G[1].

Step 4. Grasp and lift the sternum with forceps. Use scalpel to cut the pectoral girdle free from the body. This will expose the heart.

Step 5. Cut through body wall from point C to Y and then to D and D[1] as you did when the skin was cut.

Step 6. Pin back the cut body wall on both sides. Remove the thin peritoneal membrane with fine scissors if it was not removed with the cut body wall. We are now ready to study the internal organs.

Coracoid   Clavicle

Sternum

Figure 86. Cutting Through Body Wall and Pectoral Girdle of the Frog to Expose the Internal Organs.

gether. Still other membranes support the organs (see Figs. 88 and 89). Both types of membranes are called *mesenteries*. Each mesentery has a special name depending upon the organs with which it is associated. All the mesenteries are special parts of the peritoneum.

Fig. 87 shows the main organ systems and their arrangement in the male frog. The female frog differs basically from the male frog only with respect to the presence of female reproductive structures. Study page 147, "The Urogenital (Excretory and Reproductive) Systems," before you begin to dissect the female frog.

In Fig. 87 the digestive organs have been moved aside so that you can see the structures of other systems. The liver sections or lobes are shown lifted and turned toward the head to reveal the gall bladder and its tubes, and the heart. Only one lung is shown but, of course, there is another lung on the other side of the heart. With probe and forceps move the parts of your specimen so that they are in the same positions as those shown in Fig. 87. Now locate each of the labeled parts.

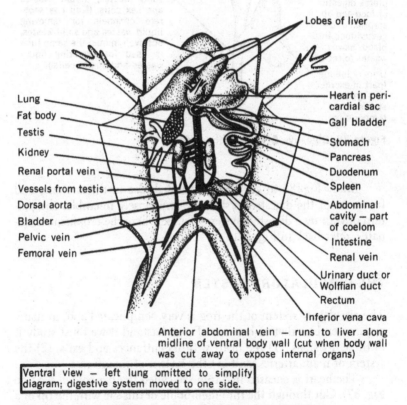

Labels (clockwise from top right):
- Lobes of liver
- Heart in pericardial sac
- Gall bladder
- Stomach
- Pancreas
- Duodenum
- Spleen
- Abdominal cavity — part of coelom
- Intestine
- Renal vein
- Urinary duct or Wolffian duct
- Rectum
- Inferior vena cava

Labels (left side):
- Lung
- Fat body
- Testis
- Kidney
- Renal portal vein
- Vessels from testis
- Dorsal aorta
- Bladder
- Pelvic vein
- Femoral vein

Anterior abdominal vein — runs to liver along midline of ventral body wall (cut when body wall was cut away to expose internal organs)

Ventral view — left lung omitted to simplify diagram; digestive system moved to one side.

Figure 87. Internal Anatomy of the Male Frog—Ventral View.

141

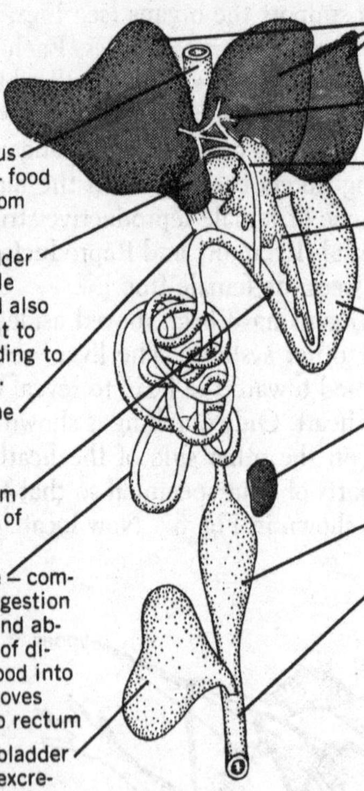

Liver — large glandular organ; produces bile juice, important in digestion of fats

Cystic duct — delivers bile juice to gall bladder

Duct in pancreas — delivers a digestive juice to intestine

Pancreas — secretes digestive juice which enters the intestine through the common bile duct

Stomach — receives food from esophagus; stores and digests food

Spleen — actually a functional part of circulatory system rather than of the digestive system

Rectum — temporarily stores, then excretes solid wastes

Cloaca — receives and expels solid wastes, liquid wastes and sex cells. (Man has separate channels for removing liquid wastes and solid wastes, but even in man the same tube is used for excreting liquid wastes and sperm cells.)

Esophagus (gullet) — food enters from mouth

Gall bladder stores bile juice and also delivers it to tube leading to intestine.

Membrane supports organs.

Duodenum first part of intestine

Intestine — completes digestion of food and absorption of digested food into blood; moves wastes to rectum

Urinary bladder (part of excretory system) stores urine temporarily.

Figure 88. Digestive System of the Frog.

Study Fig. 88 to get a more detailed idea of the anatomy and functions of the digestive system. It very closely resembles the pattern of your own digestive system. To see where the esophagus originates, refer back to Fig. 82b.

THE CIRCULATORY SYSTEM

The circulatory system of the frog is very complicated and, in many ways, resembles that of humans. To understand it we must study it in three divisions: (1) the heart with its entrances and exits, (2) the system of main arteries, and (3) the system of main veins.

The heart is encased in a thin sac called the *pericardial sac* (see Fig. 87). Cut through the thin membrane of this sac with the tip of a very sharp razor. Do not cut the heart itself. Then spread the mem-

142

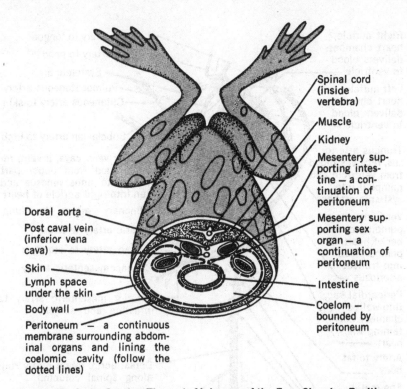

Dorsal aorta

Post caval vein (inferior vena cava)

Skin

Lymph space under the skin

Body wall

Peritoneum — a continuous membrane surrounding abdominal organs and lining the coelomic cavity (follow the dotted lines)

Spinal cord (inside vertebra)

Muscle

Kidney

Mesentery supporting intestine — a continuation of peritoneum

Mesentery supporting sex organ — a continuation of peritoneum

Intestine

Coelom — bounded by peritoneum

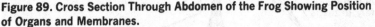

Figure 89. Cross Section Through Abdomen of the Frog Showing Position of Organs and Membranes.

brane with forceps to expose the heart. Now locate the parts of the heart as shown in Fig. 90. (This is a ventral view. Other parts of the heart will be seen when the heart is turned around for a dorsal view [see Fig. 92]). Study the arrows in Fig. 90, especially those in and near the heart. The blood vessels that carry blood away from the heart are arteries. Those that carry blood to the heart are veins. Notice that the *auricles* are fed by veins. The *ventricle* connects with one main vessel, the *truncus arteriosus*, which curves back toward the dorsal side and forms two loops around the esophagus and air tubes (see Fig. 91). These tubes are called the *systemic arches*. The two arches unite near the spine close to the stomach and form a giant artery called the *dorsal aorta*. These two arches represent a primitive vertebrate condition. In humans only one systemic arch remains.

Use probe and forceps to locate all the arteries shown in Fig. 90. A pen flashlight and a magnifying glass will be very helpful. Move the probe along each artery to see into which organ it goes. Once the arteries reach an organ they divide into smaller and smaller branches

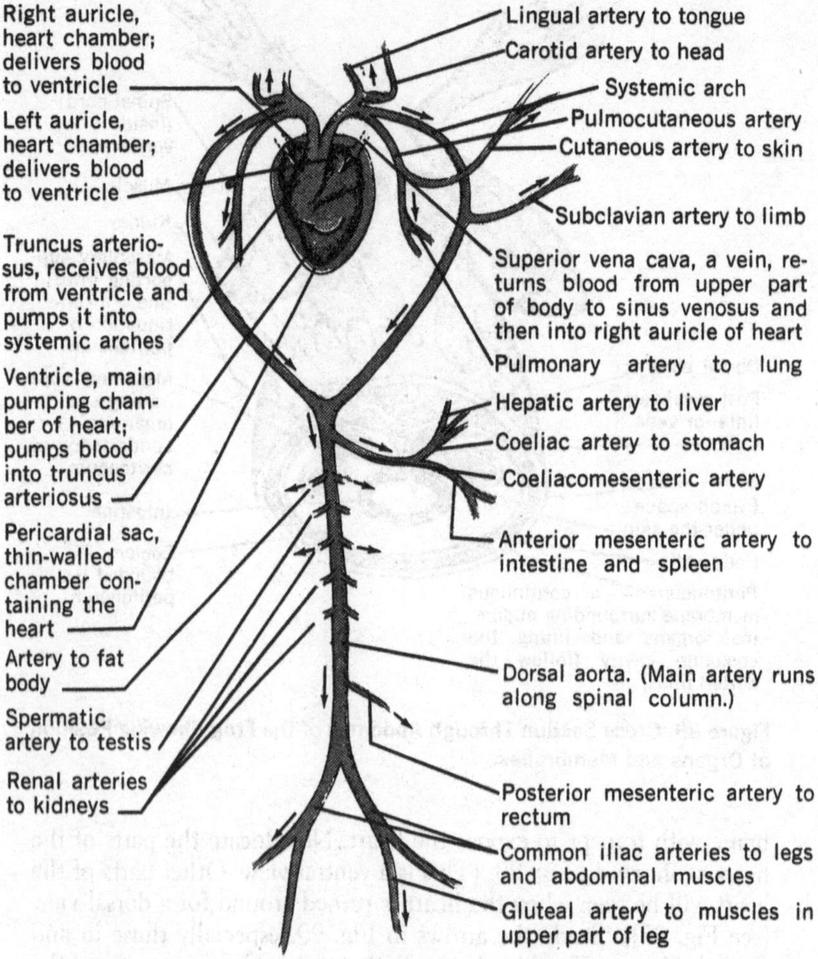
Right auricle, heart chamber; delivers blood to ventricle
Left auricle, heart chamber; delivers blood to ventricle
Truncus arteriosus, receives blood from ventricle and pumps it into systemic arches
Ventricle, main pumping chamber of heart; pumps blood into truncus arteriosus
Pericardial sac, thin-walled chamber containing the heart
Artery to fat body
Spermatic artery to testis
Renal arteries to kidneys
Lingual artery to tongue
Carotid artery to head
Systemic arch
Pulmocutaneous artery
Cutaneous artery to skin
Subclavian artery to limb
Superior vena cava, a vein, returns blood from upper part of body to sinus venosus and then into right auricle of heart
Pulmonary artery to lung
Hepatic artery to liver
Coeliac artery to stomach
Coeliacomesenteric artery
Anterior mesenteric artery to intestine and spleen
Dorsal aorta. (Main artery runs along spinal column.)
Posterior mesenteric artery to rectum
Common iliac arteries to legs and abdominal muscles
Gluteal artery to muscles in upper part of leg

Figure 90. Heart and Main Arteries of the Frog—Ventral View.

like the roots of a tree, until they form microscopic vessels called *capillaries.* The capillaries then run into larger vessels which finally emerge from the organs as veins. These veins carry blood toward larger veins (see Fig. 93) until they reach a part of the heart called the *sinus venosus* (see Fig. 92). Inside the sinus venosus is an opening to the right auricle. Through this opening passes the blood from all parts of the body except the lungs. The blood from the lungs returns to the heart directly into the left auricle. In humans there is no sinus venosus, but a small important thickening is present called the

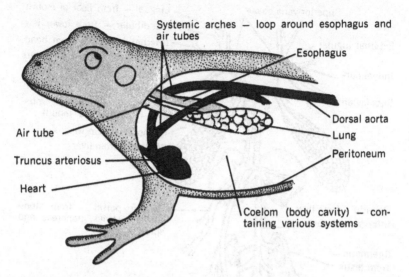

Systemic arches — loop around esophagus and air tubes

Esophagus

Dorsal aorta

Lung

Peritoneum

Air tube

Truncus arteriosus

Heart

Coelom (body cavity) — containing various systems

Figure 91. Aortic (Systemic) Arches of the Frog—Side View.

sinuauricular node which has an important job of helping to set the rhythm of the heartbeat. With Fig. 93 as a guide, trace all the main veins before you study the dorsal side of the heart.

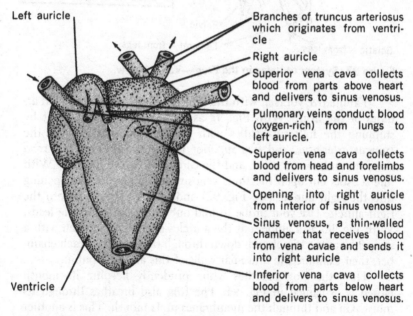

Left auricle

Branches of truncus arteriosus which originates from ventricle

Right auricle

Superior vena cava collects blood from parts above heart and delivers to sinus venosus.

Pulmonary veins conduct blood (oxygen-rich) from lungs to left auricle.

Superior vena cava collects blood from head and forelimbs and delivers to sinus venosus.

Opening into right auricle from interior of sinus venosus

Sinus venosus, a thin-walled chamber that receives blood from vena cavae and sends it into right auricle

Inferior vena cava collects blood from parts below heart and delivers to sinus venosus.

Ventricle

Figure 92. Heart of Frog—Dorsal View.

145

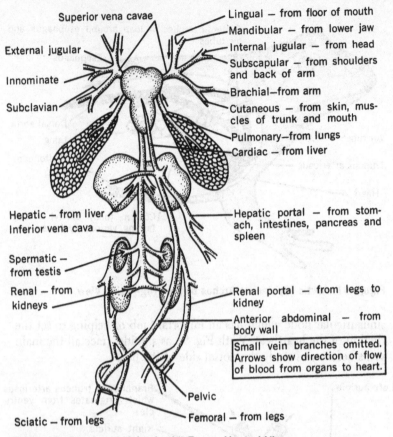

Superior vena cavae

External jugular

Innominate

Subclavian

Lingual — from floor of mouth
Mandibular — from lower jaw
Internal jugular — from head
Subscapular — from shoulders and back of arm
Brachial—from arm
Cutaneous — from skin, muscles of trunk and mouth
Pulmonary—from lungs
Cardiac — from liver

Hepatic — from liver
Inferior vena cava

Spermatic — from testis

Renal — from kidneys

Hepatic portal — from stomach, intestines, pancreas and spleen

Renal portal — from legs to kidney

Anterior abdominal — from body wall

Small vein branches omitted. Arrows show direction of flow of blood from organs to heart.

Pelvic

Sciatic — from legs

Femoral — from legs

Figure 93. System of Veins in the Frog—Ventral View.

Familiarize yourself with the location of the main veins and arteries of the heart. With Fig. 92 as a guide, remove the heart by snipping the following tubes with scissors: two branches of the *truncus arteriosus,* and two *superior vena cavae,* the *inferior vena cava,* the *pulmonary veins,* and the *cardiac vein* (see Fig. 93). With fine scissors cut open the sinus venosus at the level of the opening into the right auricle (see Fig. 92) and trace the opening into the right auricle. Use your probe to find out where each opening leads. With Fig. 90 as a guide, cut the auricles and ventricle open with a scalpel from the top of each down through the middle of each chamber, then examine the muscular walls of this amazing pump.

How the frog breathes while prudently keeping its mouth closed is explained in Fig. 94. The frog also breathes through its moist skin and through the membranes in its mouth. This is another reason why we find frogs living close to a source of water.

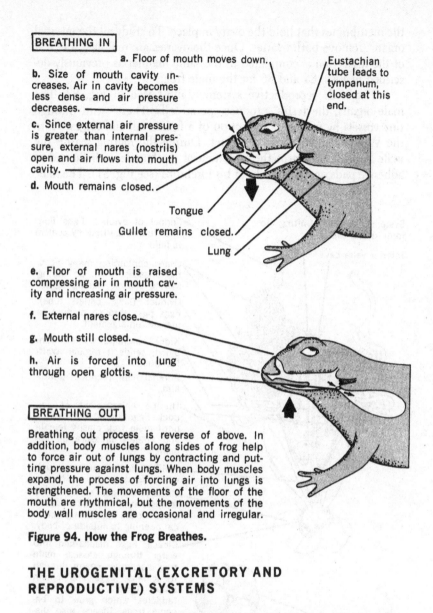

BREATHING IN

a. Floor of mouth moves down.

b. Size of mouth cavity increases. Air in cavity becomes less dense and air pressure decreases.

c. Since external air pressure is greater than internal pressure, external nares (nostrils) open and air flows into mouth cavity.

d. Mouth remains closed.

Eustachian tube leads to tympanum, closed at this end.

Tongue

Gullet remains closed.

Lung

e. Floor of mouth is raised compressing air in mouth cavity and increasing air pressure.

f. External nares close.

g. Mouth still closed.

h. Air is forced into lung through open glottis.

BREATHING OUT

Breathing out process is reverse of above. In addition, body muscles along sides of frog help to force air out of lungs by contracting and putting pressure against lungs. When body muscles expand, the process of forcing air into lungs is strengthened. The movements of the floor of the mouth are rhythmical, but the movements of the body wall muscles are occasional and irregular.

Figure 94. How the Frog Breathes.

THE UROGENITAL (EXCRETORY AND REPRODUCTIVE) SYSTEMS

Fig. 95 shows the urogenital system of the female frog. During the breeding season the entire abdomen is swollen with eggs that cover most of the organs. To study the female reproductive organs it is necessary to first remove one ovary with its eggs before we dissect the female frog. Otherwise practically nothing can be seen. The ovary can be removed easily by opening the abdomen as shown in Figs. 85 and 86, lifting the masses of eggs with forceps, and snipping with scissors

147

the membranes that hold the ovary in place. To study all the internal organs, remove both ovaries. Once the ovaries are removed, the rest of the dissection is completed in the same manner as previously described in Figs. 85 and 86 for the male frog.

The male reproductive system is illustrated in Fig. 87. The male organs, the testes, produce sperm cells. These swim through tiny vessels by means of the action of a finlike tail and migrate into the Wolffian duct, or urinary duct. During the mating season the male frog tightly clasps the abdomen of the female with the aid of adhesive pads on the thumbs of his forelimbs (see Fig. 81). The male

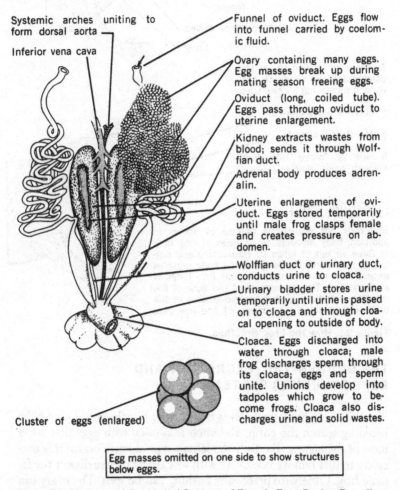

Systemic arches uniting to form dorsal aorta

Inferior vena cava

Funnel of oviduct. Eggs flow into funnel carried by coelomic fluid.

Ovary containing many eggs. Egg masses break up during mating season freeing eggs.

Oviduct (long, coiled tube). Eggs pass through oviduct to uterine enlargement.

Kidney extracts wastes from blood; sends it through Wolffian duct.

Adrenal body produces adrenalin.

Uterine enlargement of oviduct. Eggs stored temporarily until male frog clasps female and creates pressure on abdomen.

Wolffian duct or urinary duct, conducts urine to cloaca.

Urinary bladder stores urine temporarily until urine is passed on to cloaca and through cloacal opening to outside of body.

Cloaca. Eggs discharged into water through cloaca; male frog discharges sperm through its cloaca; eggs and sperm unite. Unions develop into tadpoles which grow to become frogs. Cloaca also discharges urine and solid wastes.

Cluster of eggs (enlarged)

Egg masses omitted on one side to show structures below eggs.

Figure 95. Diagram of Urogenital System of Female Frog During Breeding Season—Ventral View.

148

frog literally squeezes out the eggs from the female frog. The sperm move into the cloaca of the male frog and are then discharged over the eggs of the female frog, thus starting the life cycle anew. This clasping reflex is so overwhelmingly compelling in the male frog during the breeding season that—as proved by experiments—he will clasp a block of wood placed before him even after his head has been cut off.

IN CHARGE—THE NERVOUS SYSTEM

The nervous system consists mainly of a brain located between the roof of the mouth and the surrounding bones in the dorsal part of the head (see Figs. 96, 97, and 98). Ten pairs of *cranial nerves* go

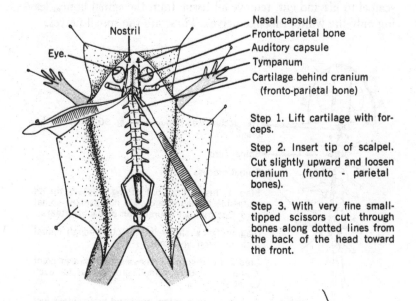

Nostril

Eye

Nasal capsule
Fronto-parietal bone
Auditory capsule
Tympanum
Cartilage behind cranium (fronto-parietal bone)

Step 1. Lift cartilage with forceps.

Step 2. Insert tip of scalpel. Cut slightly upward and loosen cranium (fronto - parietal bones).

Step 3. With very fine small-tipped scissors cut through bones along dotted lines from the back of the head toward the front.

Step 4. Press the head gently together between two fingers (one finger at point A and the other at B). This will loosen skull bones.

Step 5. With forceps pick off fronto-parietal bone and adhering tissue. The brain is now exposed.

Figure 96. Exposing the Brain and Spinal Column of the Frog—Dorsal View.

149

from the brain to the eyes, tongue, other parts of the head, and into the abdominal section. The brain is a concentrated mass of ganglia and nerves which connect with a spinal cord. The spinal cord is surrounded by the spinal column, consisting of bones called vertebrae. Ten pairs of spinal nerves reach all the parts of the body below the head. The combination of brain, spinal cord, and nerves makes up the communications network that controls the behavior of the frog.

Since the main parts of the nervous system are in the dorsal part of the frog, the first thing we must do now is to turn the frog dorsal side up on the dissecting pan. The next step is to pin the frog to the pan and cut away the skin on each side of the spinal column. (See Fig. 96.) Use the same techniques in removing the skin and body wall from the dorsal side as you did for removing the skin from the ventral side at the beginning of the dissection (see Fig. 85). Then cut through the body wall with scissors and remove the entire cut section of the body wall from tip of upper jaw to cloaca. Using forceps and scalpel to lift and cut, remove all tissue from the spinal bones, leaving only the white, tubelike nerves. These are the spinal nerves.

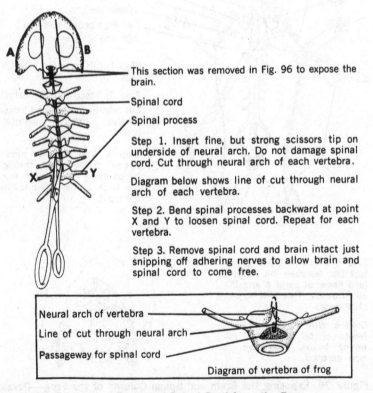

This section was removed in Fig. 96 to expose the brain.

Spinal cord

Spinal process

Step 1. Insert fine, but strong scissors tip on underside of neural arch. Do not damage spinal cord. Cut through neural arch of each vertebra.

Diagram below shows line of cut through neural arch of each vertebra.

Step 2. Bend spinal processes backward at point X and Y to loosen spinal cord. Repeat for each vertebra.

Step 3. Remove spinal cord and brain intact just snipping off adhering nerves to allow brain and spinal cord to come free.

Neural arch of vertebra

Line of cut through neural arch

Passageway for spinal cord

Diagram of vertebra of frog

Figure 97. Removal of Brain and Spinal Cord from the Frog.

150

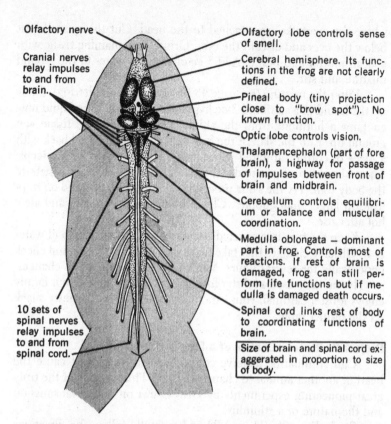

Olfactory nerve

Cranial nerves relay impulses to and from brain.

10 sets of spinal nerves relay impulses to and from spinal cord.

Olfactory lobe controls sense of smell.

Cerebral hemisphere. Its functions in the frog are not clearly defined.

Pineal body (tiny projection close to "brow spot"). No known function.

Optic lobe controls vision.

Thalamencephalon (part of fore brain), a highway for passage of impulses between front of brain and midbrain.

Cerebellum controls equilibrium or balance and muscular coordination.

Medulla oblongata – dominant part in frog. Controls most of reactions. If rest of brain is damaged, frog can still perform life functions but if medulla is damaged death occurs.

Spinal cord links rest of body to coordinating functions of brain.

Size of brain and spinal cord exaggerated in proportion to size of body.

Figure 98. Schematic Diagram of the Frog's Central Nervous System— Dorsal View.

Now follow the directions on Fig. 96 for exposing the brain and spinal column. Be very careful to avoid injuring the brain and spinal nerves. When you have completed this, go on to Fig. 97 to dissect out the brain and spinal cord. Then identify the parts of the nervous system shown in Fig. 98.

We have completed a basic dissection of the frog. Now let's apply our knowledge in some experimental projects.

Project 1: Changing the Spots on a Leopard

We have all heard the expression "You can't change the spots on a leopard." Nobody expects you to try it. But we can change the spots on a leopard frog. This, however, should be done in school under the supervision of a biology teacher.

Obtain two live, pegged small crabs or lobsters from a fish market and a live leopard frog from a biological supply house. With scalpel remove the eyestalks by cutting out a small ring of tissue where the eyestalks are attached to the head. Cut the eyestalk just below the eyes and discard the eyes. Grind the remaining tissue with a clean mortar and pestle. Add 5 cc of distilled water to the ground mixture and stir.

Filter the mixture (see page 48 for directions on filtering). Draw up the liquid in a sterilized 2-cc hypodermic syringe. The liquid now contains a hormone from the animal's sinus gland, the tissue you ground up. Rub the skin of the frog near the middle of its back with a piece of cotton saturated with alcohol. Lift the skin with forceps. Inject about 0.5 cc just under the skin of the frog. Do not penetrate the body wall. Try the same procedure on large tadpoles and on tropical fish larger than guppies. Clean the syringe with water and alcohol after use.

Keep the frog in a covered container with about 1 inch of water in it. Observe changes that take place during the first hour and check periodically for several hours. You will see some dramatic changes.

If you have used a lobster for this experiment, and if your family enjoys lobster cardinal, your experiment may end with a tasty snack on the dinner table.

Project 2: Remote Control of a Frog's Heartbeat

Can stimulating a nerve connected with one heart cause the heart of another animal to change its beat? This was one of the truly great pioneering experiments by Otto Loewi on nerve transmission and the nature of a stimulus.

Study Figs. 90, 92, and 99 and carefully follow the directions for this project. Prepare 500 ml of Ringer's solution. A good book on animal physiology will list the ingredients of Ringer's solution. Dissect out the hearts of two live frogs, anaesthetized by keeping them for 5 minutes in a 5% water–urethane solution. Leopard frogs make good subjects. Cut the blood vessels as shown in Fig. 99. Be sure to leave about 2 inches of the vagus nerve attached to one of the hearts.

Using fish line, tie one superior vena cava and the inferior vena cava of each of the two hearts. (See Figs. 90 and 92 to identify parts of heart and blood vessels.) Leave right superior vena cava open. Tie one branch of the systemic arch.

Attach funnels at appropriate levels to ring stands. Now insert each funnel end into each open superior vena cava. Tie each closed end of the systemic arch loosely to ring clamp to help support the hearts, as in Fig. 99. Then pour Ringer's solution into the top funnel. Wait until it begins to drip out of the open systemic arch.

To stimulate the vagus nerve, touch the cut end with a glass rod dipped in vinegar or acetic acid, or touch the vagus nerve with two ends of wires attached to a dry-cell battery—watch the dramatic results! It is an exciting experiment which you can convert to an open-

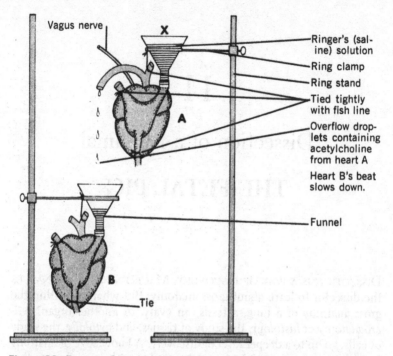

Vagus nerve

X

Ringer's (sal-
ine) solution

Ring clamp

Ring stand

Tied tightly
with fish line

A

Overflow drop-
lets containing
acetylcholine
from heart A.

Heart B's beat
slows down.

Funnel

B

Tie

Figure 99. Remote Control of the Frog's Heartbeat. (Adapted from *Animal Physiology*, by Knut Schmidt-Nielsen, © 1960 by Prentice-Hall, Inc., Englewood Cliffs, N.J.)

ended experiment by adding an additional heart to the series or by adding at point X on Fig. 99 such chemicals as adrenalin, thyroxin, oxytocin, pitocin, barbiturates, aspirin, and epsom salts.

The frog is a member of a transitional vertebrate group, the Amphibia, from which emerged the Seymouria, called by some scientists an advanced amphibian and by others a primitive reptile. There followed a long geologic period during which reptiles advanced and flourished. From prehistoric reptiles belonging to the Thecodontia, the world of birds developed. From primitive reptiles the Pelycosauria gave rise to the Therapsida, reptiles that were similar to primitive mammals. The mammalian ascent which began with reptilian ancestry has moved into the present world, after an enormous span of time. Thus, the dissection of the frog illuminates not only a basis for understanding the anatomy of mammals such as the fetal pig and man, but highlights part of the stream of dynamic biological history.

We shall now turn our attention to the dissection of a mammal, the fetal pig.

11

Dissection of a Mammal

THE FETAL PIG

DISSECTION IS A WAY OF DISCOVERY. MACRODISSECTION ENABLES the dissector to learn about gross anatomy. But what lies within the gross anatomy of a lung, a testis, an ovary, or another organ? *Microanatomy* or *histology*, the study of tissues, and *cytology*, the study of cells, go on to a deeper level of discovery. A knowledge of anatomy is made more meaningful when it is coupled with an understanding of *physiology*, the science that tries to explain how an organism functions.

The domesticated pig is an important source of food for humans. Some of its parts are used for other purposes that affect the lives of people. For example, some leather products are manufactured from the pig's skin. Our interest in the pig is focused on its excellence as a subject for dissection. However, for obvious reasons, we do not dissect an adult pig, which is a relatively large and expensive animal. Fortunately, female pigs are very fertile. They breed readily and produce fetuses which, when allowed to develop, will emerge as new generations of pigs. This capacity enables the zoologist to obtain unborn but well-developed fetuses that are very like in structure to the adult pig. There are some important differences, however, that will be noted later.

The fetal pig is an especially interesting specimen to dissect. Its dissection reveals relationships in structure with organisms lower in the evolutionary scale, like the frog for example, and enables you to examine the anatomy of a mammal whose internal parts are basically like the human anatomy. In fact, before starting this dissection it would be rewarding to review the anatomy of the frog (see Chapter 10) and to examine the human internal structure, illustrations of which are readily available in biology or anatomy textbooks.

154

TAXONOMY OF THE PIG

Every living thing has a name. There are international rules in the world of science for classifying and naming organisms. The rules belong to a branch of science called taxonomy. You should consult a textbook for the study of the principles of taxonomy.

The name of an organism consists of its genus and its species. Note that the first letter of the *genus* is always capitalized in the name, but the first letter of the *species* is not capitalized.

The scientific name of the domesticated pig is *Sus scrofa*. We will indicate the main taxonomic groups to which the *Sus scrofa* belongs, beginning with the groups that include all of the progressively more limited groups.

Fig. 100 shows the main groups of the animal kingdom arranged in an evolutionary tree of life. On page 157 is a chart showing the taxonomy of the chordates.

THE PIG IN ITS TAXONOMIC SETTING— THE MAIN GROUPS

Phylum Chordata
> Animals with bilateral symmetry, evidences of segmentation developing from three germ layers or embryonic layers
> Presence of a coelom
> Single dorsal tubular nerve cord
> Gill slits in the pharynx
> A notochord (primitive supporting rod) present at some time in the life history of the organism

Group Craniata
> Chordates with a well-defined head

Subphylum Vertebrata
> Craniates with a backbone

Superclass Tetrapoda
> Vertebrates with two pairs of limbs

Class Mammalia
> Tetrapods with hair, mammary glands, and other common features

Subclass Theria
> Marsupial (bear young in a pouch) and placental mammals

Division Eutheria
> Mammals with a placenta

Order Artiodactyla
> Hoofed, herbivorous animals with an even number of digits on each toe (cloven hooves)

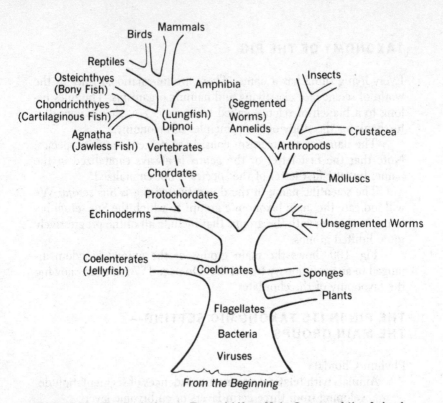

Birds
Mammals
Reptiles
Osteichthyes
(Bony Fish)
Amphibia
Insects
Chondrichthyes
(Cartilaginous Fish)
(Lungfish)
Dipnoi
(Segmented
Worms)
Annelids
Crustacea
Agnatha
(Jawless Fish)
Vertebrates
Arthropods
Chordates
Molluscs
Protochordates
Echinoderms
Unsegmented Worms
Coelenterates
(Jellyfish)
Coelomates
Sponges
Plants
Flagellates
Bacteria
Viruses

From the Beginning

Figure 100. The Evolutionary Tree of Life—Main Groups of the Animal Kingdom.

Family Suidae
> Artiodactyls with four different types of teeth in each jaw, as well as other differentiating characteristics

Genus *Sus*
> Suidae that normally interbreed

Species *scrofa*
> The domesticated pig

RATIONALE FOR FETAL PIG CLASSIFICATION

The pig belongs to the Chordates because at some time during its life history from embryo to adult it has structures typically present in the Chordates. The basic features that link the Chordates are:

> The body is bilaterally symmetrical.
> It is segmented.
> There is a coelom present.

156

Taxonomy of the Chordates

Phylum Chordata
  Subphylum Hemichordata   *Balanoglossus*
  Subphylum Urochordata (Tunicata)   *Ciona*
  Subphylum Cephalochordata   *Amphioxus*
  Subphylum Vertebrata
    Superclass Pisces
      Class Agnatha (Jawless Fishes)   *Petromyzon*
      Class Chondrichthyes (Cartilaginous Fishes)   Sharks (*Squalus*)
      Class Osteichthyes (Higher Bony Fishes)   *Perch*
    *Superclass Tetrapoda
      Class Amphibia
        Order Urodela (Caudata)   *Necturus, Salamanders*
        Order Anura (Salientia)   *Frogs and Toads*
        Order Apoda (Gymnophiona)   *Caecilians, Blindworms*
      Class Reptilia
        Order Chelonia or Testudinata   *Turtles*
        Order Squamata   *Snakes, Lizards*
        Order Crocodilia   *Crocodiles, Alligators*
      Class Aves (Birds)
      Class Mammalia
        Subclass Prototheria (Egg laying Mammals)
          Order Monotremata   *Duckbills, Spiny Anteaters*
        Subclass Theria (live bearers)
        Infraclass Metatheria (Pouched Mammals)
          Order Marsupalia   *Kangaroos, Oppossums*
        Infraclass Eutheria (Placental Mammals)
          Orders Insectivora   *Shrews, Moles*
          Order Chiroptera   *Bats*
          Primates   *Man, Apes, Monkeys*
          Carnivora   *Cats, Dogs, Tigers, Lions*
          Perissodactyla   *Horses, odd-toed Ungulates*
          Artiodactyla   *Pigs, Cattle, even-toed Ungulates*
          Proboscidea   *Elephants*
          Sirenia   *Sea Cows*
          Cetacea   *Whales*
          Edentata   *Armadillos, Sloths*
          Rodentia   *Mice, Squirrels, Rats*
          Lagomorpha   *Rabbits, Hares*

*Note: Orders are included for all Tetrapod groups except for the Aves, and at least one example is given for each group.

157

Each Chordate has a single tubular nerve cord.

The digestive system is fairly well developed.

There are paired gill slits in the pharynx.

A notochord, a cellular rod that helps to support the organs, is present.

The embryo pig has these conditions, but some become modified, reduced, or lost as the pig develops.

The pig is a Craniate (a vertebrate tetrapod) because, like other vertebrates, it has the following structures (see Fig. 102):

An internal endoskeleton, jointed skeleton, consisting of a *vertebral column* (*axial skeleton*) and a brain case (skull) or *cranium*. The axial skeleton ends in a tail.

Attached at joints to the axial skeleton are two groups of bones called girdles: the *pectoral girdle* forming the shoulder region and the *pelvic girdle* forming the hip region. To each of the girdles a pair of limbs or *appendages*, forelimbs and hindlimbs, respectively, are joined or articulated. The organism is therefore a *tetrapod*. The girdles and appendages make up the *appendicular skeleton*.

A system of muscles that attach to the bones by means of connective tissue enables the jointed skeleton to move. It also has a digestive system that has muscles.

The pig is a *mammal* because of the following characteristics:

1. hair
2. controlled body temperature
3. warm-bloodedness
4. mammary glands for suckling its young
5. two pairs of limbs
6. differentiated teeth
7. a movable tongue
8. movable eyelids
9. ears with external pinnae
10. skin with several types of glands
11. seven neck vertebrae (cervical vertebrae)
12. two occipital condyles that joint the skull to the vertebral column
13. vocal cords
14. lungs
15. a muscular diaphragm that separates the thorax from the abdominal cavity
16. a completely four-chambered heart

17. mature red blood corpuscles of a non-nucleated nature
18. urinary bladder
19. penis for sexual mating
20. internal fertilization
21. pregnant pig has a placenta, and embryonic membranes form around the developing embryo
22. a brain with twelve pairs of cranial nerves

Therefore, the pig is a member of the Eutheria.

The pig is a chordate, vertebrate mammal included in the order Artiodactyla. It belongs to this order chiefly because the third and fourth digits of each foot are about equal in structure, and a line divides these digits, forming the cloven hooves. Thus the Artiodactyla is even-toed (examples: camel, cattle, and deer). Odd-toed hoofed mammals belong to the Perissodactyla.

Family Suidae

The pig and other members of this family have four well-developed digits on each foot, with unfused metacarpal and metatarsal bones. The two outer digits on each foot do not reach the ground.

Another main reason for classifying the pig as a member of the family Suidae is the four different kinds of teeth that correspond in the upper and lower jaws. This feature separates the pig from cattle, deer, and camels, but enables the hippopotamus to be classified with the family Suidae.

Genus and Species

The *Sus scrofa* is our target for dissection and study. We have placed the pig in its family and will begin to examine the specimen itself, after a look at the basic plan of its structure.

GENERAL STRUCTURE OF A VERTEBRATE

Fig. 101 is not a diagram of any real, specific vertebrate. It does show the main parts of vertebrates in general and the approximate location of the parts as a vertebrate pattern. Of course, an animal that is a lung breather will not have gill slits in its adult life. The converse is equally true.

No attempt has been made here to show the girdles or limbs. These will be examined later in relation to the pig's skeleton. Animals with a well-developed vertebral column have only reduced notochordal remains. But the notochord is present during the early life of a vertebrate.

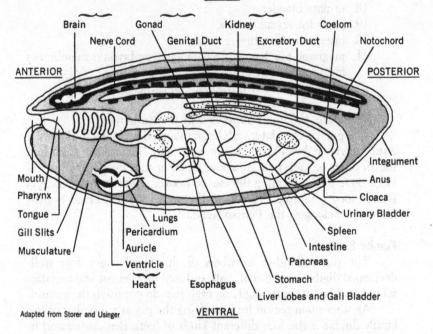

DORSAL

Brain   Gonad   Kidney   Coelom
  Nerve Cord   Genital Duct   Excretory Duct   Notochord

ANTERIOR   POSTERIOR

                                                        Integument
Mouth   Anus
Pharynx   Cloaca
Tongue   Lungs   Urinary Bladder
Gill Slits   Pericardium   Spleen
Musculature   Auricle   Intestine
                          Ventricle   Pancreas
                 Heart   Esophagus   Stomach
                                          Liver Lobes and Gall Bladder

Adapted from Storer and Usinger   VENTRAL

Figure 101. A Generalized Plan of Vertebrate Anatomy—Sagittal View, Showing Where Main Structures Are Located.

The basic framework or scaffolding of the tetrapod vertebrates, including the pig, is the internal skeleton which, except for the limbs and part of the skull, is entirely dorsally located. Much muscle under the skin or *integument* attaches to the skeleton. Other parts of the body such as the digestive organs and blood vessels also have considerable muscle tissue not connected with the skeleton.

The internal organs are located in cavities. The viscera, reproductive organs, and excretory organs are in a large cavity called the *coelom* (see Fig. 89, p. 143; also see Fig. 101 and Fig. 106). The *cloaca*, a common receptacle for wastes and sex cells, becomes reduced and is replaced, becoming the channel for waste products and for sex cells, both eggs and sperm. Instead of the cloaca the adult pig has separate openings for the discharge of solid wastes and urine, with a more complex reproductive arrangement for effecting internal fertilization and development.

Before the actual dissection is begun it is important to bear in mind that the specimen is a fetus, not a fully developed pig. The specimen was removed from the placenta of a pregnant pig before

160

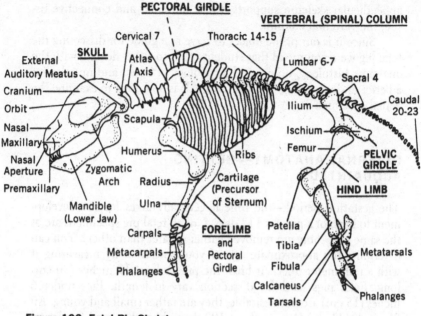

PECTORAL GIRDLE

VERTEBRAL (SPINAL) COLUMN

Cervical 7

Thoracic 14-15

SKULL

External
Auditory Meatus

Atlas
Axis

Lumbar 6-7

Sacral 4

Cranium

Orbit

Ilium

Caudal
20-23

Scapula

Nasal

Ischium

Maxillary

Humerus

Femur

Nasal
Aperture

Zygomatic
Arch

Ribs

PELVIC
GIRDLE

Premaxillary

Radius

Cartilage
(Precursor
of Sternum)

HIND LIMB

Mandible
(Lower Jaw)

Ulna

Carpals

Patella

FORELIMB
and
Pectoral
Girdle

Tibia

Metatarsals

Metacarpals

Phalanges

Fibula

Calcaneus

Phalanges

Tarsals

Figure 102. Fetal Pig Skeleton.

birth. Thus, there are some differences that are significant. The musculature of the fetal pig is difficult to dissect because the muscles are not clearly evident in their connections. In addition, the skeleton is still largely cartilaginous, not bony as in the adult. For these reasons we will not attempt to study the musculature in our dissection. However, we can utilize a diagram of the skeleton of a fetal pig and an actual skeleton of a cat and other mammalian skeletons that may be available in the laboratory. These will provide a satisfactory basis for understanding the mammalian skeleton, including man's, though there are differences in limb bones, vertebrae, and skull.

The main structural sections of the fetal pig skeleton are shown in Fig. 102. Diagrammed are the two basic skeletal divisions: the *axial skeleton,* which consists of the skull and vertebral column, and the *appendicular skeleton,* consisting of the pectoral girdle (shoulder bones), the articulated forelimbs, the pelvic girdle (hip bones), and the articulated hind limbs. Note the even number of phalanges (digits) in each limb, characteristic of the Artiodactyla which includes pigs. Consider also how excellently the axial skeleton serves as a strong support for the weight and attachment of the muscles and internal organs which depend from the vertebral column. The vertebral column and its powerful muscles hold the heavy head, and the

161

appendicular skeleton supports, by its muscular and connective tissue, the vertebral column.

Since it is our prime object to serve as a guide for dissecting the fetal pig we recommend that students refer to a text dealing with the anatomy, histology, origin, and physiology of bones and muscles for a better understanding of these tissues. It is now time to examine the actual specimen.

EXTERNAL ANATOMY—GETTING ACQUAINTED

The gestation period—the time it generally takes for the development to birth of a pig—is 112 days. Not all fetal pig specimens are of the same age; some are removed earlier or later than others. You can determine the approximate age of your specimen by measuring it with a millimeter ruler. At birth the pig is about 12 inches (30 cm) long. Fetal pigs used for dissection vary in length. Pigs under 6 inches (15 cm) are not desirable; they are rather small and young. An 8½-inch (22-cm) fetus is about 100 days old and is good for dissection. For comparison, a 1⅗-inch (40-mm) pig is about 56 days old, or about halfway through its gestation period. Naturally, the rate of development is not uniform. It does accelerate as the pig gets older, since there are more cells that multiply in later periods of gestation than in earlier periods.

Place the specimen in a dissecting pan and rinse it in tap water to remove excess preservative which may be irritating to some people. Use Figures 103 and 104 to identify external parts. Determine the sex of your specimen.

INTERNAL ANATOMY—PREPARING THE SPECIMEN FOR DISSECTION

Place the specimen ventral side up on a dissecting pan. Loop a string around a foreleg and another around a hind leg; then pass the strings under the pan to attach to the opposite legs, applying enough tension to stretch the legs apart, as shown in Fig. 105. Gently and firmly pull the umbilical cord laterally to your left. Now, with forceps, grip the skin a little less than 1 inch (2 cm) directly anterior to the site of attachment of the umbilical cord to the body wall. Lift the skin and make an incision with a sharp scalpel, just long enough to insert the blunt end of dissecting scissors. (CAUTION: Cut just deep enough

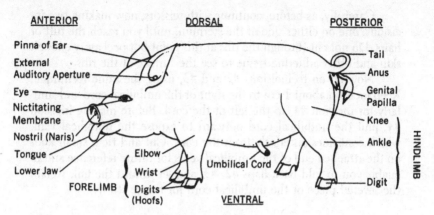

Figure 103. External View of the Fetal Pig—Female Fetus.

Labels (clockwise): ANTERIOR, DORSAL, POSTERIOR, Tail, Anus, Genital Papilla, Knee, Ankle, HINDLIMB, Digit, VENTRAL, Umbilical Cord, Digits (Hoofs), Wrist, Elbow, FORELIMB, Lower Jaw, Tongue, Nostril (Naris), Nictitating Membrane, Eye, External Auditory Aperture, Pinna of Ear

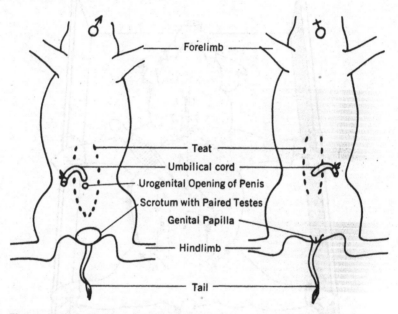

Figure 104. Ventral View of Male Fetus (left) and Female Fetus (right)—External Structures.

Labels: Forelimb, Teat, Umbilical cord, Urogenital Opening of Penis, Scrotum with Paired Testes, Genital Papilla, Hindlimb, Tail

to penetrate the body wall.) Insert scissor's end and carefully cut along median line (incision # 1). Use a round-end probe in side-to-side motions to disengage connective tissue as you progress with the scissors. With probe or your finger feel for the first obstruction where the muscular diaphragm is attached to the ribs. Do not cut the ribs or diaphragm.

163

Carefully, as before, continue with scissors, now making two incisions, one on either side of the sternum, until you reach the tuft of hairs. Do not cut through the ribs at this point but peel away enough skin and other adhering tissue to see the surface of the ribs.

Now go on to incisions #2 and #3, using the same technique. Keep incision about 1 cm to the right of the umbilical cord and similarly for incision #4, to the left of the cord. Before making incision #4, pull the umbilical cord outward to expose the umbilical vein, which continues anteriorly from the cord. Cut and tie a small knot on the attached end of the umbilical veins for future reference and to enable you to fold back flaps #2, #3, and #4 toward the tail. Retain the severed piece of the umbilical cord for future study.

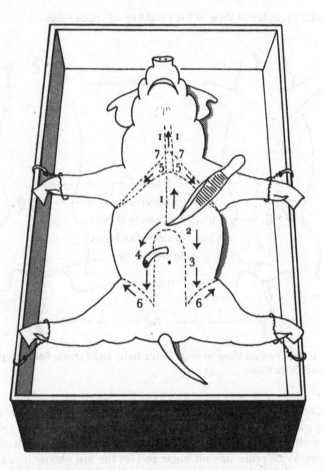

Figure 105. Exposing the Internal Organs—Order of Incisions—Ventral Position for Dissection.

Start incision #5, after determining the position of the diaphragm with probe and fingers. Without damaging the diaphragm, free it from its connection to the rib cage on both sides of the median incision. Then start and complete incision #6. Be especially careful not to cut too deeply or the penis (if it is a male specimen) may be damaged. It is fairly close to the surface and posterior to the umbilical cord.

Now wash away the excess preservative and coagulated material from the inside of the body by running tap water gently over the exposed organs to prevent damaging them. Free the incised sections below the diaphragm and, with dissecting pins, pull each section laterally toward the sides of the pan and pin the flaps securely to the paraffin bed. This will expose the abdominal cavity, with its organs.

DIGESTIVE ORGANS IN THE ABDOMINAL CAVITY (SEE FIG. 106)

A first view of the exposed visceral organs, the organs present in the abdominal cavity, reveals chiefly a glistening mass of coiled tubes—the intestines—capped by a large, dark red structure—the liver. Little else can be seen readily without manipulating the parts. The illustrations accompanying the text will show the structures in manipulated positions to emphasize the relationships of the various parts to each other. No individual parts should be severed unless so instructed or there will be damage, causing considerable difficulty in tracing the course of the circulatory system, as well as other connected structures. The student is cautioned to realize that neither anatomical illustrations nor actual photographs are perfect representations of the specimens. They are used as an aid to locating and studying the anatomy but cannot serve as a valid substitute for the real specimen; there may be variations in size and position of parts with each specimen examined.

Liver
The liver is a very large organ that has digestive and other functions. It is dark red, has several lobes or divisions and is located just posterior to the diaphragm. Raise the liver mass toward your left and locate the *gall bladder* and the *bile duct*, which conducts previously stored bile juice from the gall bladder to the small intestine.

Stomach
The stomach is on the left side, below or dorsal to part of the liver. Pull the stomach slightly posteriorly to see its connection with the esophagus, which extends from the base of the pharynx via the

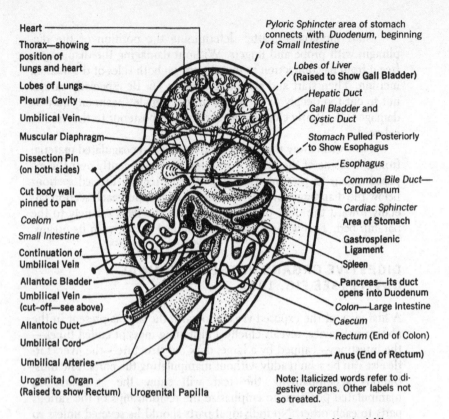

Heart

Thorax—showing position of lungs and heart

Lobes of Lungs

Pleural Cavity

Umbilical Vein

Muscular Diaphragm

Dissection Pin (on both sides)

Cut body wall pinned to pan

Coelom

Small Intestine

Continuation of Umbilical Vein

Allantoic Bladder

Umbilical Vein (cut–off—see above)

Allantoic Duct

Umbilical Cord

Umbilical Arteries

Urogenital Organ (Raised to show Rectum)   Urogenital Papilla

Pyloric Sphincter area of stomach connects with *Duodenum,* beginning of *Small Intestine*

Lobes of Liver (Raised to Show Gall Bladder)

Hepatic Duct

Gall Bladder and *Cystic Duct*

Stomach Pulled Posteriorly to Show Esophagus

Esophagus

Common Bile Duct— to Duodenum

Cardiac Sphincter Area of Stomach

Gastrosplenic Ligament

Spleen

Pancreas—its duct opens into Duodenum

Colon—Large Intestine

Caecum

Rectum (End of Colon)

Anus (End of Rectum)

Note: Italicized words refer to digestive organs. Other labels not so treated.

Figure 106. Digestive Organs of the Abdominal Cavity—Ventral View. (Thorax Exposed to Show Relationship with Abdominal Cavity—Small Intestine Pulled to Your Left and Large Intestine to Right—Coils Much Reduced.)

neck and thoracic cavity, emerging through the diaphragm to join the stomach and its cardiac end. The posterior end of the stomach is the *pylorus* where it is connected with the beginning of the small intestine. At each end of the stomach are ring-shaped muscles called *sphincters* (*cardiac* and *pyloric*) whose contractions control respectively the intake and expulsion of food.

Find the spleen, a narrow, red organ curved along the stomach. The spleen is hematopoietic (a blood cell-forming organ), not a digestive structure. It is also a center for the elimination of blood cells.

Pancreas

The pancreas is below the stomach, i.e., dorsal and posterior to the stomach. It is light in color and looks somewhat like an irregular froth of bubbles called globules. The spleen, pancreas, and stomach

166

are held together by a strong, flexible membrane of connective tissue cells called a *mesentery*. Such membranes are the principal supporting and anchoring structures that hold the internal organs in place. They also provide a pathway for blood vessels.

Locate the *pancreatic duct*, which leads from the pancreas to the first part of the small intestine called the duodenum.

Intestines

The intestines are made up of two major parts: the *small intestine*, very long, very convoluted, and rather narrow, and the *large intestine*, also convoluted but much shorter and wider. Both have strong involuntary muscles and digestive glands. The first part of the small intestine is the *duodenum*, a short tube into which the common bile duct and the pancreatic duct secrete their respective juices. The next part of the small intestine is a much longer section, the *jejunum*, followed by the *ileum*, also fairly long. However, in the pig the distinction between the jejunum and the ileum is vague. Near the end of the ileum is a pouch called the *caecum* which opens via the *ileo-caecal valve* into the large intestine, the *colon*. The last part of the colon is the *rectum*, which finally ends in a muscular aperture called the *anus*, controlled by a *sphincter muscle*. The small intestine should be moved to the left and the large intestine to the right to see relationships. The intestines should not be removed or circulatory vessels will be destroyed. However, with much patience and care, using the fine, sharp point of a scalpel, you can clear away the epithelial and connective tissue, the mesenteric membranes, that hold the intestines together and support them. The blood vessels have been injected with colored latex by the supplier, the arteries red and the veins blue. The blood vessels are also ensheathed by the *mesenteries*. By carefully removing the mesenteries without cutting the blood vessels that interpenetrate the mesenteries, the course of blood in the viscera can be traced. It will also make it easier to locate the male or female urogenital system, which lies mostly dorsal to the intestines.

The student should complement the dissection of each organ system with a review of the comparable system of the frog's anatomy and a study of the human's organ systems. Viewing dissected parts under a binocular stereomicroscope, and specimen tissue slides under a high-power microscope is highly recommended. In addition, each system should be diagrammed for better understanding of anatomical relationships and for retention of knowledge.

The Urogenital System (See Fig. 107 for Male Urogenital System, Fig. 108 for Female Urogenital System)

The urinary system and the genital system in both sexes have common receptacles and for that reason the two systems are treated together. It is advantageous at this stage for students to work in pairs, one with a male specimen and the other with a female specimen. Or, if this is not feasible, demonstration dissections showing the male and female urogenital systems should be available for examination.

Before starting the dissection of the urogenital system it would be helpful to look at an "anatomical road map" of its parts. Begin at the cut-open end of the *umbilical cord*. There are four apertures. The latex dyes immediately proclaim the *umbilical* vein and the two *umbilical arteries*. The remaining aperture is the *allantoic duct*. This leads to the expanded section, the *allantoic bladder*. At the attached end of the umbilical cord are two tubes that join the posterior of the bladder, the *ureters*. At the end of the bladder is a broader tube, the *urethra*, which in the adult male pig continues into and through the *penis*. Its opening is posterior to the point of attachment of the umbilical cord to the body.

In the female the urethra opens directly into the *urogenital*

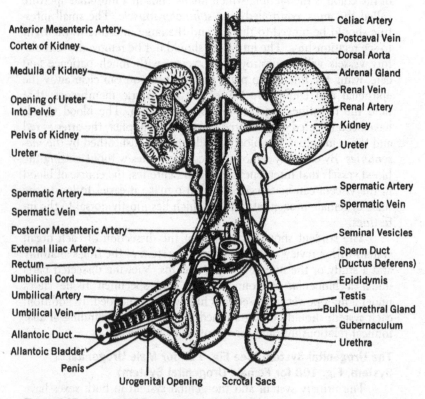

Figure 107. Urogenital System of the Male Pig.

168

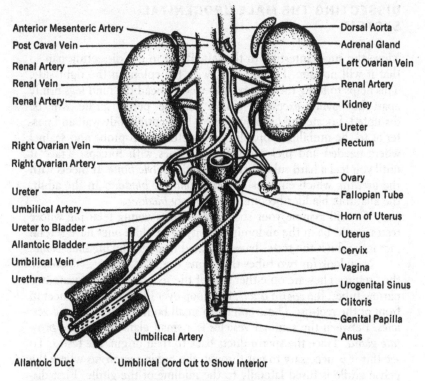

Anterior Mesenteric Artery

Post Caval Vein

Renal Artery

Renal Vein

Renal Artery

Right Ovarian Vein

Right Ovarian Artery

Ureter

Umbilical Artery

Ureter to Bladder

Allantoic Bladder

Umbilical Vein

Urethra

Allantoic Duct

Umbilical Artery

Umbilical Cord Cut to Show Interior

Dorsal Aorta

Adrenal Gland

Left Ovarian Vein

Renal Artery

Kidney

Ureter

Rectum

Ovary

Fallopian Tube

Horn of Uterus

Uterus

Cervix

Vagina

Urogenital Sinus

Clitoris

Genital Papilla

Anus

Figure 108. Urogenital System of the Female Pig.

sinus. In both the male and female pig the ureters meet the kidneys at the *hilum,* the site of the inward curve of the kidneys, and conduct liquid wastes from the kidneys to the urethra.

The male organs that produce sperm are the *testes.* In young fetal pigs the testes are in the abdominal cavity. In older ones and in the adult the testes have migrated into *scrotal sacs* on a connective tissue cord called the *gubernaculum* through two openings in the abdominal cavity, the *inguinal canals.* The sperm cells pass from each testis through the coiled *epididymis,* then through the sperm duct, and finally out through the urethra in the penis.

In the female, a pair of *ovaries* lies posterior to the kidneys. Coiled tubes (the *oviducts* or *Fallopian Tubes*) continue to the curved *horn of the uterus* on each side of the body. At the posterior end of the uterus is a constriction called the *cervix,* immediately below which is the *vagina,* which is joined by the *urethra* in a common receptacle called the *urogenital sinus.* The final aperture is the *urogenital pore* with its *urogenital papilla* and the *clitoris.*

169

DISSECTING THE MALE UROGENITAL SYSTEM

Move the small intestine to the left. Pin it to the dissecting pan so that it will not slip. Do the same with the colon on the right side. This helps to bring the *kidneys* into view. Spread the hind legs firmly apart but avoid excessive tension or the pelvic girdle will be disrupted. Locate the *penis* just below the ventral body wall and posterior to the umbilical cord. Free it carefully with probe and scalpel where needed and pick away fringe tissues with forceps. Trace it until you feel a hard substance. This is the *pelvic bone.* It meets with the *urethra,* which empties into the *allantoic bladder.* In the adult, the allantoic bladder becomes the *urinary bladder.*

We will assume your specimen is not a young fetal pig whose testes would be in the abdominal cavity. The older ones have scrotal sacs into which the testes have migrated through the *inguinal canal.*

Now look for two tubes that come from the kidneys. These are the ureters. They are on either side of the rectum. There, locate two narrow tubes, the *sperm ducts,* that loop over the ureters and meet in front of the rectum. These enter two small bodies called *seminal vesicles.* Between the seminal vesicles is a small gland called the *prostate gland.* Trace the sperm ducts back to their origin, the testes. To see this it is necessary to cut through the *pubic symphysis* where the pelvic girdle is fused laterally to the midline of the girdle. First dissect away the tissues of the body wall. Beneath this there are muscles which must be removed until you can see a pale line. Touch this to confirm its hardness. This is the pubic symphysis. With the point of a sharp scalpel pierce the line to divide the girdle. DO NOT CUT DEEPLY.

It is necessary to cut open the inguinal canal to follow the sperm ducts into the scrotal sacs, which also should be slit open. Find the epididymis, a tightly coiled mass on the surface of each testis. Each sperm duct is a continuation of the epididymis. The sperm cells produced by the testes pass through the epididymis, the sperm ducts, the seminal vesicles, and the urethral part of the penis to effect internal fertilization. Finally, locate the paired *spermatic arteries* and *veins* that lead into the scrotal sacs.

DISSECTING THE FEMALE UROGENITAL SYSTEM

Identify the bean-shaped kidneys. At the apex of each kidney is an *adrenal gland,* an endocrine gland of great physiological importance that is not a urogenital organ. A little caudad (toward the tail) of the

170

kidneys are the *ovaries*, light-colored, egg-shaped structures bounded by a fringe of tiny tubules, the *Fallopian tubes*. From the ovary trace the Fallopian tubes to the horns and body of the *uterus*. Continue identification of the *cervix* and *vagina*. As explained previously, the vagina and urethra join to form the *urogenital sinus*.

Repeat the procedure for cutting the pubic symphysis, as explained above. Find the confluence of the vagina and the urethra. This is where the urogenital sinus begins. Trace the sinus to the *urogenital pore*, the *urogenital papilla* and the *clitoris* just below it.

DISSECTING THE RENAL PART OF THE UROGENITAL SYSTEM

While dissecting the genital system of the male and female fetal pig, several of the urinary structures were observed and their association with the genital systems were noted. We can therefore start with the ureters. Find the place where the ureters connect with the inward curve of the kidneys. Strip away the *peritoneal membrane* from one kidney and free it, without cutting the connecting blood vessels. Identify the *renal veins* coming from the *postcaval vein* and the *renal arteries* arising from the *dorsal aorta*, which is alongside the postcaval vein.

Carefully slice the kidney sagitally, that is, on a longitudinal and equal split along the middle plane (like separating the two halves of a peanut). With a hand lens or a binocular microscope, observe the open spaces in the sliced kidney halves. These spaces, called the *pelvis*, form a continuous drainage system from the *renal tubules* to the *ureters* and successively to the *bladder*, *urethra*, *urogenital sinus* and *urogenital pore*. Although it is not sharply clear in the fetal condition, you may observe that the kidneys have two distinct areas, the inner dark area called the *medulla* and a lighter section around the medulla called the *cortex*, with its tubules and rich blood supply.

EXAMINING THE INTERNAL ANATOMY OF THE THORAX (SEE FIGS. 109 AND 112)

Complete incision #7 using bone scissors (see Fig. 105) to cut through the ribs on both sides of the *thorax*. Be careful not to cut too deeply or you may damage the lungs and heart. Remove the cut sections to expose the *pleural lining* of the lungs and the *pericardial sac* encasing the heart. The *diaphragm* was freed when you performed incision #5.

Locate the lobed *lungs*, the *heart*, and part of the *thymus*. The

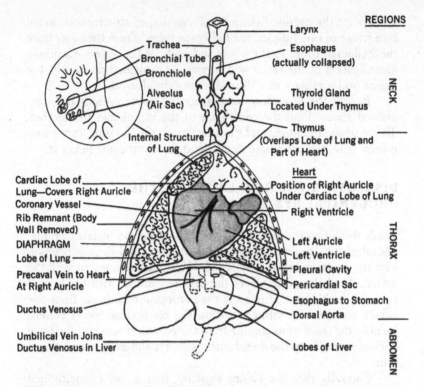

Larynx

Trachea
Bronchial Tube
Bronchiole

Esophagus
(actually collapsed)

Alveolus
(Air Sac)

Thyroid Gland
Located Under Thymus

Internal Structure
of Lung

Thymus
(Overlaps Lobe of Lung and
Part of Heart)

Heart

Cardiac Lobe of
Lung—Covers Right Auricle

Coronary Vessel

Rib Remnant (Body
Wall Removed)

DIAPHRAGM

Lobe of Lung

Precaval Vein to Heart
At Right Auricle

Ductus Venosus

Umbilical Vein Joins
Ductus Venosus in Liver

Position of Right Auricle
Under Cardiac Lobe of Lung

Right Ventricle

Left Auricle
Left Ventricle
Pleural Cavity
Pericardial Sac
Esophagus to Stomach
Dorsal Aorta

Lobes of Liver

Figure 109. Organs of Respiratory System in Thorax and Neck—Ventral View. Other Organs Represented to Show Relative Positions.

cavity which is almost obliterated by the lungs and heart is the *thoracic cavity*. It is part of the larger cavity called the *coelom*. The organs posterior to the diaphragm are in the *abdominal coelom* and the *peritoneal cavity*, while the heart and lungs are in the *thoracic coelom*, all of which are lined with *epithelial cells*.

There are actually three divisions of the coelom in the thorax. Between the left and right lung the heart lies in its *pericardial cavity*. In turn, the pericardial cavity, thymus, part of the dorsal aorta, and the precaval vein are in the *mediastinal cavity*, which is an almost completely filled space. It extends medially from the sternum and dorsally to the vertebral column, containing blood vessels, the trachea, the esophagus, and the pericardial sac.

Pull and lift the left lung to see the dorsal aorta, esophagus, and the *phrenic nerves* that serve the diaphragm. The internal structure of the lungs with the arborization of the *bronchial tubes* leading to the microscopic *air sacs* or *alveoli* can best be seen in specially prepared sections for microscopic examination.

Delicately cut open the thin *pericardial membrane* around the heart, preparatory to a study of the circulatory system. This sequence will enable you to see the connections between the prethoracic and postthoracic parts of the digestive, respiratory, and circulatory systems. However, before going forward with the dissection of the circulatory system we will dissect part of the head and neck.

INTERNAL STRUCTURES OF THE HEAD AND NECK (SEE FIG. 110)

Starting at the upper jaw near the proximal end of the snout (the end attached to the body), cut through the skin toward the ear and below the eye. Continue the incision to the neck and back, and forward to the base of the snout. Remove the skin to expose superficial structures. Curving over and around a large muscular mass called the *masseter* are branches of the *facial nerve*. The nerve is white. Along the posterior curve of the facial nerve is the large, reddish *parotid gland*, one of the *salivary glands*. Trace the *parotid duct* from the gland ventrally and below the masseter to the mouth. Below the parotid is another salivary gland, the *submandibular* or *submaxillary gland*. Its duct opens to the floor of the mouth. Another salivary gland, the *sublingual*, lies forward and ventral to the submandibular gland. Its duct, which is difficult to trace, also opens into the floor of the mouth. Flanking the parotid gland are the branches of the *facial*

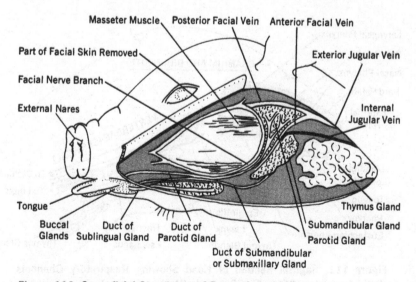

Figure 110. Superficial Structures of Face—Lateral View.

173

vein which arise from the *external jugular vein*. The *internal jugular* may be seen near the thymus gland.

INVESTIGATING THE INTERIOR OF THE MOUTH AND PHARYNX (SEE FIG. 111)

Use the upper surface of the tongue as a guide for freeing the upper jaw from the lower jaw. With scissors, cut laterally through the external wall. Gradually spread the mouth open. With your finger explore the roof of the mouth to feel the forward *hard palate* and the posterior *soft palate*. Beyond the end of the soft palate is the *epiglottis*, a valvelike flap at the opening of the *larynx*. Just above the epiglottis at the *laryngeal pharynx* is the beginning of the *esophagus*, which you previously met in the thorax. The esophagus connects with the stomach, after it penetrates the diaphragm.

The main structures of the digestive organs in the mouth are lips, teeth, tongue, jaws, pharynx, hard and soft palates, salivary glands, and esophagus. The adult pig has four types of teeth: *incisors, canines, premolars,* and *molars.*

The dental formula of the fetal pig is written as I ⅓, C¼, P¼, and M%. The fractions refer to the teeth of the upper jaw and lower jaw. Since the molars do not appear in the fetal pig the formula for fetal molars is M%.

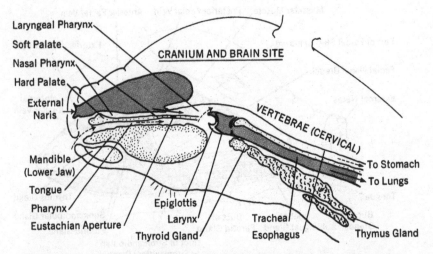

Figure 111. Sagittal Section of Head Showing Respiratory Channels (Solid Arrows) and Digestive Channels (Dashed Arrows).

174

Your specimen probably has some erupted incisors. Remove one from the lower jaw for examination.

THE CIRCULATORY SYSTEM (STUDY FIGS. 112 THROUGH 118)

A living organism, plant or animal, unicellular or multicellular, is dependent upon its external environment for life-sustaining chemicals and sources of direct or auxiliary energy. All parts of the organism selectively utilize these needed chemicals. So it is that even in microscopic unicellular organisms a mode of transport is evident. Whether it is by a constant streaming of protoplasm or a network of minute canals, the distribution of chemicals is accomplished and the organism lives. In multicellular life an internal transport system is highly organized and structuralized. Such a system requires a flowing medium, the blood and the lymph. A propelling structure, the heart, with its muscular tissue, a closed circuit of tubes, arteries, and veins, and an interconnecting network of fine vessels, the capillaries, all serve the cells of the body. Thus the blood and lymph, in "making the rounds," manage to service all the cells of the organism unless a blockage occurs, caused either by disease or accident. The circulatory system is therefore not merely a pump with tubes attached; it is alive and its function goes beyond mechanical transport. Blood and lymph are also major factors in combating disease and in temperature control.

Let us now examine the circulatory system of the fetal pig and see how it differs from the circulatory system of the adult pig.

THE PATHS OF BLOOD IN THE FETAL PIG—PLACENTA TO PLACENTA

It is important to remember that the fetal circulation starts with the fetal placenta. While soluble food and oxygen pass from the maternal placenta into the blood stream of the fetus, fetal blood and fetal blood cells do not. They are independently formed and remain within the circulatory system of the fetus. The structural circulatory system of the fetus is "weaned away" from the mother, starting with its first breath and producing several changes associated with the development of the lungs as the new respiratory mechanism. When the fetus is born, the umbilical "lifeline" is parted from the mother and an independent life system is ready for growth.

175

Let us first examine the main paths of the fetal plan of circulation and then see how it differs from the adult plan of circulation.

THE PLACENTA-TO-PLACENTA CIRCUIT

1. Oxygen- and food-rich blood from the *fetal placenta* pass through the *umbilical vein.*
2. The blood continues to the *ductus venosus* at the liver, connecting with the *postcaval vein.* Some blood is diverted through *liver sinuses* into the *hepatic veins,* which also join the postcaval vein.
3. Blood continues to the *right atrium* of the heart, through the *foramen ovale* (an aperture between the right and left atria) and into the *left atrium.* The foramen ovale serves as a pulmonary bypass.
4. From the left atrium the blood is forced into the *left ventricle.*
5. The blood, still enriched, is then pumped into the aorta and via arterial branches to the head, shoulder, and forelimbs— *not* to the lungs.
6. Veins conduct blood with diminished oxygen from the head and forelimbs through the precaval vein to the right atrium.
7. From the right atrium the blood moves into the *right ventricle.*
8. The right ventricle drives blood into *pulmonary arteries* to the lungs and, via the *ductus arteriosus,* into the *descending aorta;* then to the hind limbs and visceral organs (see Fig. 115).
9. Finally, deoxygenated blood returns to the *placenta* through the paired *umbilical arteries* carrying metabolic wastes, to renew the food and oxygen supply and to remove cellular wastes.

ADULT CIRCULATION—HEART TO HEART

1. The right division of the heart may be considered the beginning of the blood circuit. The *right atrium* receives deoxygenated blood from the *precaval vein* and the *postcaval vein* (see Figs. 112 and 113).
2. Blood is propelled into the *right ventricle.* The *foramen ovale* is closed, preventing passage of blood from the right atrium to the left atrium.

3. The *right ventricle* pumps blood into the lungs via the *pulmonary arteries*. No blood enters the *ductus arteriosus* because that vessel becomes a closed cord after birth.
4. The blood is oxygenated in the lungs, returning to the left atrium of the heart through the *pulmonary veins*. This is the first of two major heart-to-heart circuits, the *pulmonary circuit* and the *systemic circuit*.
5. From the left atrium blood is forced into the left ventricle, which has powerful cardiac muscles.
6. The contractions of the left ventricle create a strong propulsive action that drives the oxygenated blood to all the rest of the body, excluding the lungs. Part of the original placental circuit becomes converted into nonconducting structures; the *ductus venosus*, the *umbilical vein*, and the *umbilical arteries* remain as *ligaments*.
7. Blood is returned from the capillaries all over the body, excluding the lungs, which by successive confluences form ever larger tributary veins, ultimately leading into the very large precaval and postcaval veins. The caval veins deliver deoxygenated blood to the right atrium from the rest of the nonrespiratory parts of the body. This begins the second major heart-to-heart circuit, the *systemic circuit*.

TRACKING THE VENOUS CIRCULATION (SEE FIGS. 112 AND 113)

A logical place to start is at the heart. You have already partially removed the pericardial membrane exposing the ventral surface of the heart. Using Fig. 112 as a reference source, locate the right and left *auricles*, flaplike structures covering the right and left *atria* or chambers, which are at the anterior end of the heart. Posterior to the auricles are the right and left *ventricles*. Bear in mind that the pig is on its back so that its left side faces your right side. Observe the large vessel that descends to meet the right auricle. This is the *precava* or precaval vein.

We will temporarily postpone dissection of the heart until we have tracked the main venous and arterial vessels. It is important to remember that the venous system of the fetal pig cannot be represented by a fixed blueprint. Some of the vessels may not appear individually but are joined with others, and are not always found at the expected position. However, it is still possible to locate the main veins and arteries, particularly since they are well color-coded, red for arteries and blue for veins.

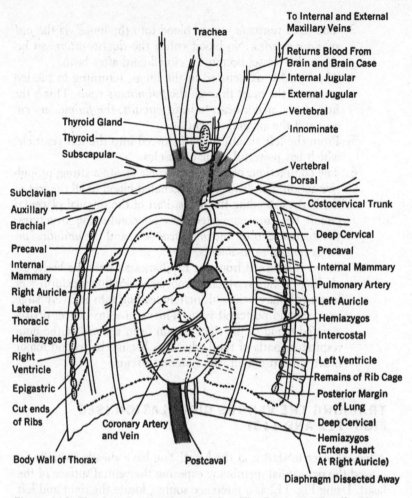

Trachea

To Internal and External
Maxillary Veins

Returns Blood From
Brain and Brain Case

Internal Jugular

External Jugular

Vertebral

Innominate

Thyroid Gland

Thyroid

Subscapular

Vertebral

Dorsal

Costocervical Trunk

Subclavian

Auxillary

Brachial

Deep Cervical

Precaval

Precaval

Internal
Mammary

Internal Mammary

Pulmonary Artery

Right Auricle

Left Auricle

Lateral
Thoracic

Hemiazygos

Hemiazygos

Intercostal

Right
Ventricle

Left Ventricle

Epigastric

Remains of Rib Cage

Posterior Margin
of Lung

Cut ends
of Ribs

Coronary Artery
and Vein

Deep Cervical

Body Wall of Thorax

Postcaval

Hemiazygos
(Enters Heart
At Right Auricle)

Diaphragm Dissected Away

Figure 112. The Anterior Venous Vessels—The Precaval "Tree"—Ventral View.

Patiently clear away tissue from the precaval vein and its anterior and lateral branches. Remove most of the thymus gland and thyroid gland. Slide the probe under and along blood vessels to free them. Muscle masses should be raised, freed, and removed, always being watchful not to injure the larger blood vessels and nerves, which look like light nylon strands.

The first branch you will meet, moving anteriorly, is on the left side. It is the *costocervical trunk.* Find its three branches, *dorsal, deep cervical,* and *vertebral.* Some of the *intercostal veins* from the thorax also return blood to the costocervical trunk.

The precaval vein divides to form two short trunks called the *in-*

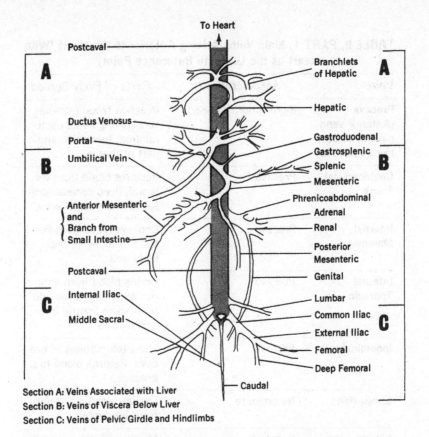

To Heart

Postcaval

A

Branchlets of Hepatic

A

Hepatic

Ductus Venosus

Portal

Umbilical Vein

B

Gastroduodenal

Gastrosplenic

Splenic

Mesenteric

Phrenicoabdominal

Adrenal

Renal

B

Anterior Mesenteric and Branch from Small Intestine

Posterior Mesenteric

Postcaval

Genital

Internal Iliac

C

Middle Sacral

Lumbar

Common Iliac

External Iliac

Femoral

Deep Femoral

C

Caudal

Section A: Veins Associated with Liver
Section B: Veins of Viscera Below Liver
Section C: Veins of Pelvic Girdle and Hindlimbs

Figure 113. The Posterior Venous Vessels—The Postcaval "Tree."

nominate veins. Branches fan out from the innominates on the right and left sides. The main ones, in addition to the costocervical veins shown in Fig. 112, are drawn somewhat extended to enable the dissector to see more readily the sequence of branches of the precaval vein.

Referring to Table II, Part 1 (p. 180), find these paired veins in the following order:

1. Costocervical trunk
2. Internal mammary
3. Subclavian and its continuations, Axillary and Brachial
4. Subscapular
5. Internal jugular
6. External jugular

Now go on to the dissection of the veins posterior to the heart. (Refer to Table II, Part 2, and to Fig. 113).

TABLE II, PART 1. Main Veins Arising Anterior to the Heart (With Heart as the Ultimate Reference Point)

Veins	Associated with	Parts of Body Served
Precava (Anterior vena cava or superior vena cava)	Right atrium of heart	Receives blood from vessels serving head, pectoral area, forelimbs, and part of thorax
Costocervical Trunk	Precava	Receives blood from vertebral, deep cervical, dorsal, and subcostal veins
Internal Mammary	Precava	Conveys blood from ventral ribs along ventral body wall
Internal Thoracic	Precava	Drains blood from anterior abdominal area and from some ventral intercostal veins
Innominate	Precava	Short bifurcations of precava. Returns blood to precava
Subclavian	Innominate	Carries blood returning from forelimbs
Lateral Thoracic	Subclavian	Muscles of thoracic wall (mainly)
Subscapular	Near union of axillary and subclavian	Shoulder
External Jugular	Innominate near heart and union of internal and external maxillary veins, anteriorly	Mainly muscles and other tissues of head
Internal Jugular	Innominate	Cranium and brain
Hemiazygos (single vein)	Right auricle	Transports blood from intercostal cells, cells surrounding and between the ribs, to the heart
Pulmonary	From lungs to left atrium. To locate these veins trace from left lung to heart	Returns blood from lungs to heart

180

TABLE II, PART 2. Main Veins of the Postcaval Route (With Heart as the Ultimate Reference Point)

Veins	Associated with	Parts of Body Served
Postcaval (Posterior or inferior vena cava)	Right atrium	Traverses length of body posteriorly from pelvic area to right atrium. Collects and transports blood from its tributaries
Veins of Hepatic Portal Sub-Route	Veins from liver to postcaval. Union of gastrosplenic and mesenteric veins	Deliver blood from liver to postcaval; carry blood from digestive organs to liver sinuses
Hepatic	Several veins that arise from extensive liver capillaries and blood sinuses and empty into postcaval	Forms anterior path to postcaval starting with portal vein
Umbilical	From umbilical cord to ductus venosus, a wide tube emptying into the postcaval	Conducts blood from placenta to postcaval
Renal	From kidneys, singly or possibly doubly. Runs into postcaval	Carries blood from kidneys to postcaval
Adrenal	From adrenal usually directly to postcaval. Sometimes one runs into renal vein	Adrenal glands
Genital	Enter postcaval caudad to renals (internal spermatic in male and utero-ovarian in female). Left internal spermatic may join left renal	Delivers blood from gonads to postcaval
Anterior Mesenteric	From small intestine joins mesenteric	Distributes blood from small intestine to mesenteric vein

Veins	Associated with	Parts of Body Served
Gastrosplenic	Mesenteric	Stomach and spleen
Splenic	Mesenteric	Spleen
Mesenteric	Combines with gastro-splenic and connects to form portal vein	Carries blood from stom-ach, spleen, and intes-tines to portal vein
Posterior Mesenteric	Runs into mesenteric	Blood moves from colon (large intestine) to mes-enteric vein
Lumbar	Postcaval (chiefly)	Returns blood from dor-sal interior cells of lumbar region
Common Iliac	Begins at terminal bi-furcation of postcaval	Directs blood from pelvic area and hind limbs to postcaval vein
External Iliac	Main tributary of com-mon iliac	Delivers blood to common iliacs from pelvic area and hind limbs
Middle Sacral	Thin prolongation of postcaval	Serves tissues in sacral region
Caudal	Prolongation of middle sacral	Carries blood from tail to middle sacral

Manipulate organs by lifting or moving them to the side in order to find veins that are not otherwise readily observable. Do the same when locating and identifying the arteries. Slice through and pick away membranes surrounding the blood vessels. Dissection of the heart will follow investigation of the arterial system.

THE ARTERIAL SYSTEM—SOME BASIC FACTS TO REMEMBER

Arteries are muscular and help the heart's driving action. They con-duct blood away from the heart and toward the biological consum-ers, the *cells* of the body, by way of the middlemen, the capillaries. Another circulatory middleman is the *lymphatic system*. Unlike the circulatory system, which is a *closed vascular system*, the lymphatic

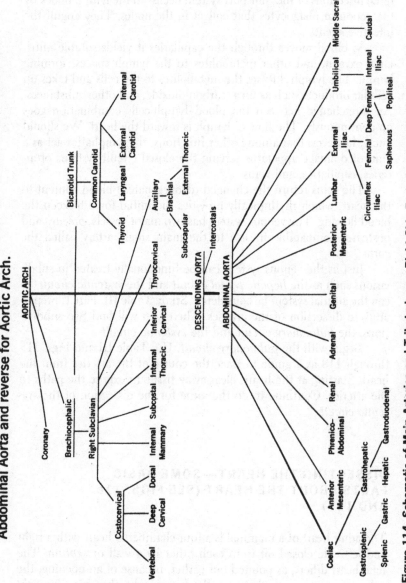

Sequence represents arterial connections mainly from cephalic to caudal axis for Abdominal Aorta and reverse for Aortic Arch.

Figure 114. Schematic of Main Arteries and Tributaries on Right Side Exclusive of Pulmonary Circuit.

183

system begins with open *tissue spaces*. These connect with *lymph capillaries*, which in turn run into larger vessels called *lymphatic ducts*. The *ducts* pass through *lymph nodes*, centers for the production of lymphocytes, on the way to the heart. Protection against bacterial infections of the transport system occurs in the lymph nodes by the action of phagocytes that collect in the nodes. They engulf the infective agents.

As blood moves through the capillaries it yields soluble nutrients, oxygen, and other metabolites to the lymph spaces, forming lymph. The lymph diffuses the metabolites to the cells and takes up cellular products such as urea, carbon dioxide, and other substances. An interchange between the blood–lymph-cells combination goes on continuously. The flow of lymph is toward the heart. We should know, however, that among other functions, the lymphatics act as a kind of drainage apparatus serving the closed circuit systems of arteries, capillaries, and veins.

The veins return the changed (deoxygenated) blood content to the heart, which rhythmically provides the initial force to keep the blood flowing. The venous system has two major vessels, *precava* and *postcava*, comparable in size to the main single artery called the *aorta*.

Just as the venous system can be functionally treated in subdivisions such as the *hepatic portal circuit* and the *systemic circuit*, so can the arterial system be considered. Study Table III, Part 1, preparatory to dissection of the arteries. There you will find two subdivisions, the *pulmonary circuit* and the *systemic circuit*.

Begin with the *pulmonary circuit*. Use Table III and Figs. 115 through 118 as a guide to trace the course of the arteries from the heart. Starting at the heart, clear away tissue to expose the paths to the arteries. Continue to do the same for the dissection of the *systemic circuit*.

DISSECTING THE HEART—SOME BASIC FACTS ABOUT THE HEART (SEE FIGS. 116 AND 117)

The adult heart of a mammal is a four-chambered heart with a right and left side closed off from each other by a wall or *septum*. The fetal heart differs, as pointed out earlier, because of an opening, the valved *foramen ovale*, between the upper right chamber, the *right atrium*, and the upper left chamber, the *left atrium*. This temporary condition is succeeded by the separateness of the right and left sides of the heart at birth.

The heart's four chambers are essentially divided into two reception and two propulsion chambers. The receiving chambers are the *left* and *right atria*, each enclosed exteriorly by a thinly muscled structure called an *auricle*. The shape is likened to a dog's ear and therefore called "auricle," derived from the Latin word for "ear."

TABLE III, PART 1. The Arterial Delivery System: Main Routes (The Pulmonary Circuit)

Arteries	Connections	Blood Delivered to
Pulmonary	Right ventricle with dichotomy forming left and right pulmonary artery branches, one to each lung	Lungs. Blood returned by pulmonary veins to left auricle
Ductus Arteriosus	A blood shunt between pulmonary artery and aorta at arch of aorta. At birth the lumen of the ductus arteriosus closes to become the ligamentum arteriosum. A dangerous congenital anomaly is sometimes found in human babies resulting from retention of open ductus arteriosum	Fetal pig: Blood is short-circuited from right ventricle into aorta. A lesser volume of blood goes into pulmonary arteries to lungs In the adult: All the blood flows from right ventricle via pulmonary artery directly to the lungs

TABLE III, PART 2. The Arterial Delivery System (The Systemic Circuit)

Arteries	Connections	Blood Delivered to
Aorta (Ascending)	From left ventricle	Aids in forcing blood beyond left ventricle
Aortic Arch	Aorta curves over anterior ventral part of heart to descend dorsally near vertebral column with branches to all parts of body except lungs (see ductus arteriosus)	Branches going to head, neck, forelimbs, and thorax

185

TABLE III, PART 2 (Cont'd.)

Arteries	Connections	Blood Delivered to
Tributaries of Aortic Arch		
Coronary arteries	Right and left coronaries arise from aorta. Right is dorsal and left lies obliquely along ventral side of heart. Right coronary joins hemiazygos	Heart cells
Brachiocephalic	Branches into the bicarotic trunk or into two separate common carotids and right subclavian	Anterior part of body from thorax to head, via its branches
Left subclavian	Arises directly from aortic arch. (Right subclavian arises from brachiocephalic)	Delivers blood to branch arteries of thorax, left forelimb, neck and head
Tributaries of Brachiocephalic	Either divides into common carotids or carotids may arise independently	(See bicarotid trunk below)
Right subclavian	Continues to forelimb as brachial artery with several branches, before becoming brachial	Note: Arterial branches of left subclavian parallel those of right subclavian
Branches of Subclavian		
Costocervical trunk	Divides into three branches	(See below)
Vertebral		Cervical vertebrae and skull
Deep cervical	Forms first intercostal	Dorsal muscles of part of neck
Dorsal		Interior muscles of upper part of back and dorsal muscles of lower part of neck

186

TABLE III, PART 2 (Cont'd.)

Arteries	Connections	Blood Delivered to
Internal mammary	Subclavian	Mammary glands, thymus and some pectoral muscles
Subcostal	Subclavian	Several intercostals
Internal thoracic	Subclavian—also branches to the superior epigastric	Ventral intercostals and parts of upper abdomen
Inferior cervical	Subclavian, before it becomes the brachial	Sides of neck and proximal part of shoulder
Thyro-cervical	Left subclavian near origin of internal mammary	Muscles of pectoral regions, thyroid and parotids
Axillary	Continuation of subclavian to beginning of forelimb, where it becomes the brachial, with subscapular and external thoracic offshoots. Leads to radial and ulnar arteries	Forelimbs
Brachial branches	From axillary	Branches of brachial
External or lateral thoracic	From brachial near subscapular	Pectoral and forelimb muscles
Subscapular	Large blood vessel near junction of axillary and brachial	Shoulder area
Bicarotid Trunk and Subsidiaries	From brachiocephalic	
Common carotid arteries	From bicarotid trunk	
Left common carotid		(See branches of left common carotid, Column 1)
Thyroid artery	From left common carotid	Thyroid gland

TABLE III, PART 2 (Cont'd.)

Arteries	Connections	Blood Delivered to
External carotid	From common carotid	Tongue and part of face
Internal carotid	Branch of common carotid	Brain (Note: several arteries intervene, which we will not track)
Descending Aorta		
Thorax: intercostals	Aorta from aortic arch to abdominal region	Ribs and body walls
Abdominal Aorta		
Coeliac	Separates to form gastrosplenic and gastrohepatic	Stomach, pancreas, spleen and liver
Gastrosplenic	Divides into gastric and splenic	
Gastric	From gastrosplenic	Stomach
Splenic	From gastrosplenic	Spleen
Gastrohepatic	Separates into gastroduodenal and hepatic	
Gastroduodenal	From gastrohepatic	Duodenum and stomach
Hepatic	From gastrohepatic	Liver
Anterior mesenteric	From aorta	Small intestine and anterior part of colon
Phrenicoabdominal	From aorta under kidney	Supplies lateral wall of body and diaphragm
Renal	Aorta to kidney	Kidney
Genital	In male: Aorta to testes via spermatic arteries	Testes in male
	In female: Aorta to utero-ovarian arteries	Uterus in female
Posterior mesenteric	Aorta to colon	Descending colon

188

Arteries	Connections	Blood Delivered to
Lumbar	Aorta (6 pairs present in lumbar region. One pair arises from middle sacral region)	Muscles and skin of parts of abdominal wall
External iliac	From aorta ramifies to form several branches	Pelvic region and hind limbs via branches
Circumflex iliac	Branch of external iliac	Abdominal wall and sides
Femoral	Branch of external iliac	Anterior part of leg
Deep femoral	Branch of femoral at division of external iliac	Thigh
Popliteal	Branch of deep femoral	Knee region
Saphenous	Branch of deep femoral	Posterior of hind limb
Umbilical	Aorta bifurcates, forming two umbilicals; continues and becomes middle sacral	To placenta. Sends blood to internal iliacs
Internal iliac	Internal iliacs arise from umbilicals	Pelvic area
Middle sacral	Aorta	Sacral region
Caudal	Branch of middle sacral	Tail

The propulsion chambers are the *left* and *right ventricles*, strongly powered with *cardiac muscle*, a special type of muscle tissue found in the heart. The left ventricle is more heavily muscled than the right ventricle.

Surrounding the heart is a thin membranous sac, the *pericardium*, consisting of two layers, the *serous pericardium* or *epicardium*, and the *fibrous sac*, with the serous layer surrounding the outside of the heart. The pericardium is a moist, slippery membrane, minimizing friction as the heart performs its contractions and relaxations.

A shallow furrow on the surface of the heart, the *coronary sulcus*, appears to form a geographical boundary between the auricles and ventricles. Coronary vessels lie in the *coronary sulcus*. The upper

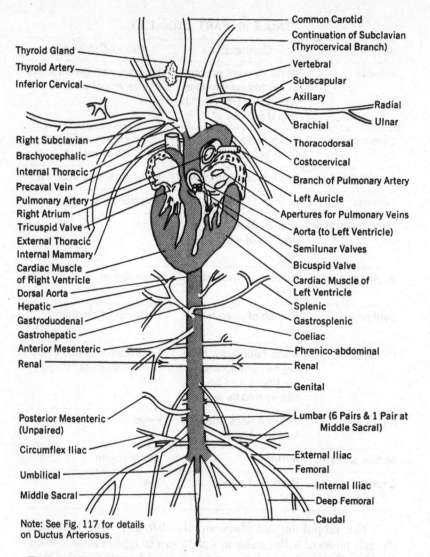

Thyroid Gland
Thyroid Artery
Inferior Cervical

Right Subclavian
Brachyocephalic
Internal Thoracic
Precaval Vein
Pulmonary Artery
Right Atrium
Tricuspid Valve
External Thoracic
Internal Mammary
Cardiac Muscle
of Right Ventricle
Dorsal Aorta
Hepatic
Gastroduodenal
Gastrohepatic
Anterior Mesenteric
Renal

Posterior Mesenteric
(Unpaired)

Circumflex Iliac

Umbilical

Middle Sacral

Common Carotid
Continuation of Subclavian
(Thyrocervical Branch)
Vertebral
Subscapular
Axillary
Radial
Ulnar
Brachial
Thoracodorsal
Costocervical
Branch of Pulmonary Artery
Left Auricle
Apertures for Pulmonary Veins
Aorta (to Left Ventricle)
Semilunar Valves
Bicuspid Valve
Cardiac Muscle of
Left Ventricle
Splenic
Gastrosplenic
Coeliac
Phrenico-abdominal
Renal
Genital
Lumbar (6 Pairs & 1 Pair at
Middle Sacral)
External Iliac
Femoral
Internal Iliac
Deep Femoral
Caudal

Note: See Fig. 117 for details
on Ductus Arteriosus.

Figure 115. Arterial System—Heart Hemisected from Base to Apex Along Midsection—Dorsal Aspect Shown.

part of the heart is called the *base;* the lower, tapered part is the *apex.*

The fetal heart is quite small, making identification of internal structures troublesome. It would be desirable to obtain a fresh sheep or cow heart for dissection, particularly since the anatomy of these adult hearts is compatible with the anatomy of the human heart.

190

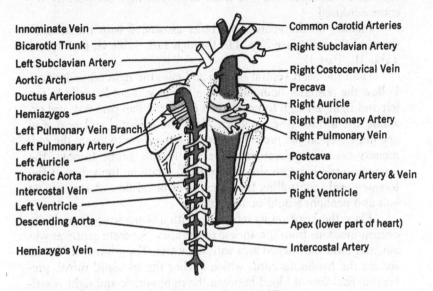

Innominate Vein —
Bicarotid Trunk —
Left Subclavian Artery —
Aortic Arch —
Ductus Arteriosus —
Hemiazygos —
Left Pulmonary Vein Branch —
Left Pulmonary Artery —
Left Auricle —
Thoracic Aorta —
Intercostal Vein —
Left Ventricle —
Descending Aorta —
Hemiazygos Vein —

Common Carotid Arteries
Right Subclavian Artery
Right Costocervical Vein
Precava
Right Auricle
Right Pulmonary Artery
Right Pulmonary Vein
Postcava
Right Coronary Artery & Vein
Right Ventricle
Apex (lower part of heart)
Intercostal Artery

Figure 116. Dorsal Aspect of Heart—External View.

DISSECTION PROCEDURES

First, free the fetal heart from its vascular connections. Start by lifting the apex and locate the postcaval vein, aorta, esophagus, and trachea. Use scissors to cut the blood vessels. Do not cut the esophagus and trachea. Then, moving to the base (anterior) of the heart, cut through the anterior vessels, freeing the heart. Cut through the aorta, very close to its site of emergence from the heart. Be absolutely sure to identify the ventral surface of the heart with a pin or toothpick,

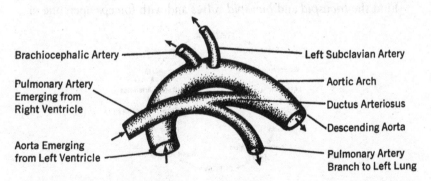

Brachiocephalic Artery —
Pulmonary Artery Emerging from Right Ventricle —
Aorta Emerging from Left Ventricle —

Left Subclavian Artery
Aortic Arch
Ductus Arteriosus
Descending Aorta
Pulmonary Artery Branch to Left Lung

Figure 117. Connections of Ductus Arteriosus—Heart Turned on Side with Left Side Facing Reader.

191

because in manipulating the freed heart you may temporarily become confused.

Identify all the veins and arteries associated with the anterior (base) of the heart (see Figs. 114 through 118, Table II, Part 1, and Table III, Part 1).

Hold the heart ventral side up and raise the right auricular flap. Follow the posterior boundary of the flap as you cut laterally to the left and right, enough to enable you to lift the flap part way and examine the atrium (the interior). Note the *precaval* and *postcaval* openings. Repeat the procedure for the *left auricle* to locate the *pulmonary vessels'* connections. Direct a flexible probe through the *foramen ovale* between the two auricles. Examine the valves at the foramen ovale controlling the direction of blood flow. A magnifying lens and penlight would be helpful.

Place the heart on its left side. With a sharp scalpel, gradually deepen incisions from the auricle to the apex. Separate gently as you cut, and flush the opened area with tap water. The whitish cords you see are the *tendinous cords* which anchor the *tricuspid valves*, preventing backflow of blood between the right auricle and right ventricle.

Repeat the lateral incisions on the opposite side until the cords appear, and flush this side with water. Locate the *bicuspid valves*. Carefully, beginning at the apex, cut left and right until the ventral half of the heart begins to separate from its dorsal half. Avoid cutting the tendinous cords. Continue until the *semilunar valves of the aorta* (left ventricle) and the *semilunar valves of the pulmonary artery* (right ventricle) appear. Continue separating the halves until they are freed from each other.

Note the muscular thickness of the *left ventricle*. It has to provide the major force for driving the blood through the *systemic circuit* which serves all of the body except the lungs. Locate the *septum*, the wall separating the right and left halves of the heart. Find the *tricuspid* and *bicuspid valves* and with forceps open one of

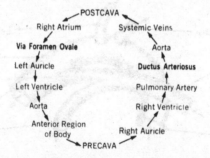

Figure 118. Schematic of Fetal Pig Heart-to-Heart Circuit (Emphasis on Foramen Ovale and Ductus Arteriosus).

the flaps and try to raise it anteriorly. Observe the controlling action of the *tendinous cords* and the *papillary muscles* to which the cords are posteriorly attached. Identify the openings of the *pulmonary artery* and *aorta*. Observe the *semilunar valves,* cup-shaped to prevent backflow of blood.

Give thought to the effects of inoperative valves, hardened by deposits of calcium and cholesterol, and inadequate valves, too small to close off the reverse leakage of blood. These conditions unfortunately do occur. Another pathological condition, resulting in *blue baby,* develops when the *ductus arteriosus* fails to close at birth, starving the infant of its needed supply of oxygen to the tissues.

Observe that the "open" space in the ventricles seems very limited. (See Fig. 115). Yet, because of its remarkable muscles, responding to hormones, neural stimuli, bioelectrical stimulation, and a variety of other factors, the heart is capable of an amazing output in an organism's life span. It is not only the quantity of output that is remarkable but the ability of the heart to respond to the moment-to-moment blood needs of the body. Run for the bus and your heart beats more rapidly. Rest, and it slows down. Its beat is always rhythmical except when diseased—a magnificent supporter of life. Protect it!

BRIEF OVERVIEW OF THE MAMMALIAN BRAIN

The nervous system is a complex coordination of cells specialized to conduct impulses throughout the organism. It enables the animal to respond to changes in its external and internal environment by receiving stimuli through *sensory nerve cells.* This serves to initiate responses of adjustment by *effectors,* generally *muscles* or *glands.* The higher forms of animal life have concentrations of *association neurons* in a special organ, the *brain,* and especially in the *cerebrum* of the brain. The presence of neuron junctions called *synapses* makes it possible for the end of one neuron to transfer its impulse to one or more of several different neurons which, at their synaptic junctions, may increase greatly the potential of variety in the response to the original stimulus. The culminating effect of potential for variation in response is *intelligent behavior.* An organism that can have but one predictable behavioral response for a specific stimulus reacts by *reflex* or *instinct.* One that can produce a number of different responses to a specific stimulus performs with complex behavior to the level of intelligence.

The nervous system is an integrated complex, but for purposes of study it is considered to have two major divisions, the *central nervous system* and the *peripheral nervous system.* The former includes the *brain* and *spinal cord.* The peripheral system consists of the *cra-*

nial nerves, the *spinal nerves' ganglia,* and organs that relate directly with the peripheral system.

The *peripheral nervous system* includes the *autonomic nervous system,* which has neural control over the involuntary muscles and glands of the viscera as well as influencing blood-vessel and heart action. The *autonomic nervous system* is subdivided into the *sympathetic* and *parasympathetic systems.* These two subdivisions generally act antagonistically. If one tends to stimulate secretion by a gland, the other tends to inhibit secretion—a neural system of checks and balances. The autonomic system extends from the base of the skull to the tail. *Preganglionic nerve fibers* terminate in *autonomic ganglia* from spinal nerves and certain cranial nerves. *Postganglionic fibers* begin at the ganglia and terminate at glands or involuntary muscles.

Parasympathetic fibers are connected with several cranial nerves and with the sacral section of the spinal cord and its spinal nerves. An *autonomic nerve center* is called a *plexus.*

THE CENTRAL NERVOUS SYSTEM: THE BRAIN

The brain is a highly organized concentration of nerve fiber tracts, nerve cell bodies, and association neurons, protectively covered by membranes and a bony cranium and nourished by blood and cerebrospinal fluid.

Divisions of the Brain (See Figs. 119 through 121)

1. Prosencephalon or *forebrain* has two divisions: *Telencephalon,* front section of the brain, consists of *cerebral hemispheres* and *olfactory lobes.*
   Diencephalon, composed of nerve transmission tracts, connects the *cerebral hemispheres* with the *midbrain.* Its *thalamic structures* form such parts as the *pineal body, posterior lobe of the pituitary gland,* and the *optic chiasma,* where fibers of the optic nerves cross over (see Figs. 119 through 121).
2. *Mesencephalon* or *midbrain* consists of the *crura cerebri,* tracts bridging the *pons* and *cerebellum* with the *forebrain.* It also has the *corpora quadrigemina,* which are centers for sight and hearing.
3. *Myelencephalon* or *brain stem* essentially consists of the *medulla oblongata,* the junction of the spinal cord and the brain. It is a continuous passageway between the brain and the spinal cord. The so-called "vital reflexes," such as sneez-

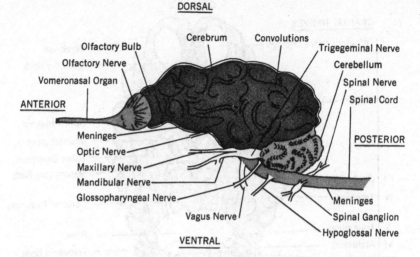

Figure 119. External Features of Fetal Pig Brain—Left Side of Brain.

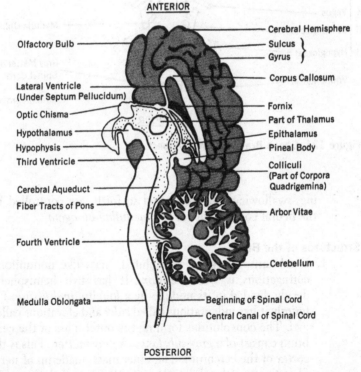

Figure 120. Sheep Brain Hemisected—Right Side with Internal Half Facing Up.

195

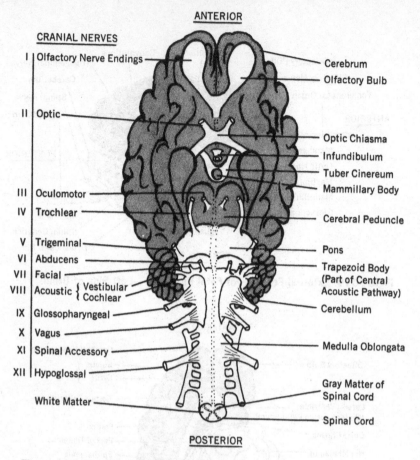

ANTERIOR

CRANIAL NERVES

I | Olfactory Nerve Endings —————————————— Cerebrum

————————————— Olfactory Bulb

II | Optic ——————————————————————— Optic Chiasma

—————————— Infundibulum

—————————— Tuber Cinereum

—————————— Mammillary Body

III | Oculomotor ———————————————— Cerebral Peduncle

IV | Trochlear ——————————————————

V | Trigeminal —————————————————— Pons

VI | Abducens ——————————————————— Trapezoid Body
(Part of Central
VII | Facial —————————————————————— Acoustic Pathway)

VIII | Acoustic { Vestibular
Cochlear —————————— Cerebellum

IX | Glossopharyngeal ——————————————

X | Vagus ——————————————————————

XI | Spinal Accessory ————————————— Medulla Oblongata

XII | Hypoglossal ——————————————————

Gray Matter of
White Matter ————————————————— Spinal Cord

————————————— Spinal Cord

POSTERIOR

Figure 121. Sheep Brain—Ventral Aspect.

ing, swallowing, breathing, and so forth, are controlled by
the neural components of the *medulla oblongata.*

Structures of the Brain

1. Cerebrum: The large, convoluted, mass like nonuniform
corrugations is the *cerebrum.* It has two hemispheres
deeply divided in its midline by a *longitudinal fissure.* Exteriorly it has indentations called *sulci* and elevations called
gyri. The convolutions forming the outer mass of the cerebrum consist of a *gray substance* or *gray matter.* This is the
cortex of the cerebrum. The inner mass, made up of nerve
fiber tracts with white fatty sheaths, is called the *white
matter.* Specific areas of the cerebrum control definite

196

functions such as sight, hearing, and memory. It is the site of intelligence.

2. Olfactory lobes: Situated at the most anterior end of the cerebrum are two large prominences, the *olfactory lobes*, that relate to the sense of smell.

3. Corpus callosum: This is a central, white commissure, a band of nerve fiber tracts interlinking the cerebral hemispheres. It lies deep between the two hemispheres.

4. Cerebellum: The cerebellum is a large structure consisting of a right and left hemisphere connected by a narrow, middle bridge. It is located between the posterior, occipital portion of the cerebrum and the brain stem, the *medulla oblongata*. The *cerebellum* connects with the *pons, midbrain* and *brain stem*. Among other functions, it is a locus for control of balance and muscular coordination.

5. Pons: The *pons* is composed of masses of nerve fibers arching over the brain stem ventrally, like a thick band. Anteriorly, it rests against the *cerebral peduncles* of the *midbrain*. Several cranial nerves arise at the pons.

6. Corpora quadrigemina: The *corpora quadrigemina* are four moundlike elevations called *colliculi*, seen ventrally by spreading apart the *cerebral hemispheres* and the *cerebellar hemispheres*. The *trochlear nerves* emerge from the corpora quadrigemina.

7. Thalamus: The *thalamus* looks like two large ovoid gray masses resting aslant, across the cephalic end of the midbrain and on each side of the *third ventricle*. It is a synaptic "relay station" to the cerebral cortex. It controls perception of pain and extremes of heat and cold.

8. Hypothalamus: The *hypothalamus* is located at the ventral and posterior region of the *Diencephalon* at the third ventricle. It comprises the following structures:

    Mammillary bodies: two small mounds near the *cerebral peduncles* at the midline.

    Tuber cinereum: located between the optic chiasma and the mammillary bodies (see Fig. 121). The *infundibulum*, a narrow, hollow stalk, arises from the tuber cinereum and terminates ventrally in the posterior lobe of the *pituitary gland*, the *hypophysis cerebri*.

    Optic chiasma: an X-shaped crisscross of nerve fibers connecting the *optic nerves* with the *lateral geniculate bodies*, near the tuber cinereum.

9. Fornix: Two bands of fibers connected in the middle under the *corpus callosum* and above the *thalamus*. Its lateral legs are the *crura* which, fused above, form the *fornix*.

10. Medulla oblongata: The caudal end of the brain, forming a transitional passage between the spinal cord and the brain. The anterior of the *medulla oblongata* is at the pons. The *central canal* of the spinal cord continues in the medulla oblongata and widens in triangular shape at its forward half, enlarging the passage into the *fourth ventricle* (see Figures 120 and 121). The *cortex* of the medulla is made up of *white matter*, while the inner mass of nerve cells is *gray matter*—the reverse of the cerebrum. The medulla oblongata connects with the cerebellum via extensions called the *restiform bodies.* It is the major center for vital reflexes.
11. Meninges: Three membranes that envelop the brain and spinal cord. The outer, relatively thick membrane is the *dura mater*; the middle is the *arachnoid*, and the innermost is the *pia mater.* The last two are very thin membranes.

Spinal Cord (See Figs. 119 and 122)

The spinal cord is a neural "superhighway" linking the brain with all parts below the medulla oblongata. It is a slender rod solidly filled with neural tissue, except for a hollow central canal. It extends from the tail to the medulla oblongata, with which it is continuous. The cord's divisions are identical with the divisions of the vertebrae

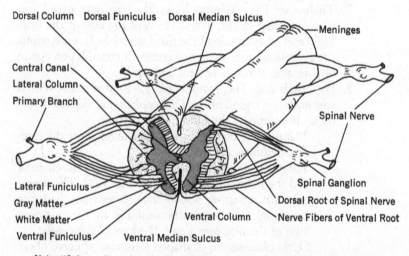

Dorsal Column   Dorsal Funiculus   Dorsal Median Sulcus

Meninges

Central Canal
Lateral Column
Primary Branch

Spinal Nerve

Lateral Funiculus
Gray Matter
White Matter
Ventral Funiculus

Spinal Ganglion
Dorsal Root of Spinal Nerve
Ventral Column   Nerve Fibers of Ventral Root
Ventral Median Sulcus

Note: "Columns" are found in Gray Matter. "Funiculi are found in White Matter. Meninges are outer covers of Spinal Cord.

Figure 122. Spinal Cord Segment. (The Schematic Diagram Shows Segmentally Paired Spinal Nerves and Cross Section of Cord with Gray Matter Projected.)

that surround it. A pair of spinal nerves emerge segmentally from the cord. The human spinal cord has thirty-one segmental pairs. The spinal cord is covered by *meninges*. The outer mass of the cord is white matter. The inner mass, shaped grossly like a butterfly with outstretched wings, contains the gray matter. Study Fig. 122 for some details of the inner structure of the spinal cord.

The dorsal part of the cord receives impulses from outside the cord and transmits them by nerve tracts to the brain or other parts, including the viscera, via the plexuses of the autonomic system. Some impulses are conducted to motor neurons located in the same spinal nerve in its ventral arm, to innervate effectors, forming a *reflex arc*. This results in a *reflex act*.

Peripheral Nervous System—Spinal and Cranial Nerves

The pairs of *spinal nerves* have *receptors* which receive stimuli at all parts of the body posterior to the brain, transmitting impulses to the spinal cord and distributing impulses through their ventral arms, to the body. Connecting with the spinal nerves are sympathetic and parasympathetic nerve centers. Such a center is a plexus. From each plexus nerves arborize to the viscera affecting the action of cardiac and smooth muscle tissue and glands.

Cranial nerves arise from the brain. There are twelve pairs of cranial nerves, numbered sequentially in Roman numerals, generally with the first beginning at the anterior end of the brain and terminating with the medulla oblongata.

Study Table IV and Fig. 121, as you identify the cranial nerves.

TABLE IV. Cranial Nerves

No.	Name	Brain Connection	Distal Connection
0.	Nervus terminalis	Cerebral hemispheres	Nasal septum
I.	Olfactory	Mucus membrane of olfactory organ	Olfactory lobe and olfactory bulb
II.	Optic	Diencephalon	Retina of eye
III.	Oculomotor	Oculomotor nucleus of mesencephalon	Several eye muscles
IV.	Trochlear	Medulla oblongata: anterior part of medullary velum	Superior oblique muscle of eye

TABLE IV (Cont'd.)

No.	Name	Brain Connection	Distal Connection
V.	Trigeminal	Myelencephalon-gasserian ganglion	Epithelium of skin, mouth and jaw muscles
VI.	Abducens	Myelencephalon	Lateral rectus muscle
VII.	Facial	Myelencephalon (See Figure 110, page 173)	Taste buds in forward part of tongue Sublingual and submaxillary glands Superficial scalp and face muscles (chin to mastoid) Platysma, muscles of ventral-lateral neck; posterior of digastric Stylohyoid muscle
VIII.	Acoustic	Myelencephalon vestibular ganglion Spiral ganglion	Semicircular canals, utriculus and sacculus of ear Organ of Corti in ear
IX.	Glossopharyngeal	Myelencephalon	Pharynx and its muscles Parotid gland Rear of tongue
X.	Vagus	Myelencephalon	Viscera of thorax and abdomen through sympathetic ganglia Voluntary muscles of pharynx and larynx Pharynx, larynx, and trachea External ear
XI.	Spinal accessory	Myelencephalon	Voluntary muscles of pharynx and larynx Via sympathetic plexus to thoracic and abdominal viscera Dorsal anterior back muscles
XII.	Hypoglossal	Myelencephalon	Ventral neck muscles Muscles of tongue

DISSECTION PROCEDURES FOR THE
NERVOUS SYSTEM

Part I. Working with the Fetal Pig

Set the specimen dorsal side up. With sharp scalpel start an incision at the midline, anterior to the eyes. Continue the incision to a midline point at the level of the anterior attachment of the ears. Cut laterally on each side, to the ears. Continue the incision anteriorly on each side but do not cut around the eyes. Complete the incisions from the left and right sides anteriorly meeting at the midline.

Lift the skin and cut more deeply, following the line of initial incisions, and gradually remove the skin, muscles, and connective tissue until the skull is exposed. Now insert the point of the scalpel as a wedge at the edge of the posterior median cut, and begin to cut through the skull. Use forceps and scalpel to pry up pieces of the skull and break them off, until you have cleared the entire exposed dome. Then the *meninges,* the *dura mater, arachnoid,* and *pia mater* membranes should be removed. Scissors will probably be most effective. At this point sever the head from the body, using your scalpel and ringing the incisions deeper and deeper until the spinal column is parted.

You will first observe the cerebrum and its hemispheres, the gyri and sulci, and the fissures (see Figs. 119 and 120). Posterior to the cerebrum find the cerebellum and its hemispheres. The fetal brain is relatively small and soft and readily damaged. It is not an effective structure for dissection and study.

Examine Fig. 119 and try to locate the parts listed in the illustration. Do not remove the brain from the cranial cavity because you will most likely destroy the few cranial connections that might otherwise be identifiable. Move the brain gently to the right or left to locate the cranial nerves shown in Fig. 119.

Take the severed torso and place it dorsal side up on your tray. Make a midline incision all the way to the tail. Cut down to the spine. Cut away the skin and muscles about ½ inch (1.8 cm) to the right and left of the midline, exposing the spinal column. Cut away two or three vertebrae to find the spinal cord which passes through the spinal column. Find the *filum terminale,* a threadlike tapered end of the spinal cord. Try to trace a *plexus,* part of the *autonomic nervous system,* at the cervical or lumbar sections of the spinal cord. By following the path of several spinal nerves, remove the skin and muscles until you discover a network of nerves. They look like white cords. This is difficult, requiring skill and patience.

Except for studying the gross anatomy of the eye and ear, your dissection of the fetal pig is finished. For continued study of the central nervous system it is desirable to obtain a sheep brain, which is large enough and whose texture and consistency is firm enough for successful dissection. If sheep brain specimens are available for examination only, not for dissection, it is possible to identify all of the main structures apparent without cutting into the brain itself. You should concurrently examine a dismemberable model of a human brain or an actual brain for purposes of comparison and for firsthand knowledge of the structure of the human brain.

Part II. Sheep Brain

Using Figs. 120 and 121 as your pictorial guide and the description of the anatomy of the sheep brain as your textual guide, start with an examination of the dorsal brain. Locate the structures in Fig. 120. Gently spread the cerebral hemispheres along the deep longitudinal central fissure to find the corpus callosum and its neighboring structures, such as the fornix, the optic chiasma, and so forth. Then proceed to spread the cerebral and the cerebellar hemispheres apart to find such structures as the pons, corpora quadrigemina, and fourth ventricle. After locating all the illustrated parts, turn the brain ventral side up and again locate all the structures discussed and diagrammed. Systematically locate and identify all twelve cranial nerves shown in Fig. 121 and in Table IV.

If a dissectable brain is available, make a sagittal section by cutting down dorso-ventrally between the cerebral and cerebellar hemispheres until they are separated. Again, using Figs. 120 and 121 as your guide, identify all the parts, continuously referring to the textual descriptions. Finally, cut a longitudinal section of the cerebrum antero-posteriorly above the level of the corpus callosum to see the interior of the cerebrum. Find the cortex and its gray matter of association neurons and nerve cell bodies and the inner mass of white matter, composed mainly of nerve tracts, with nerves encased by white myelenated (fatty) sheaths. Also cut a cerebellar hemisphere longitudinally in half to observe its internal structure of arbor vitae and nerve tracts.

Dissecting the Eye (See Figs. 123 and 124)

Carefully slit all the skin at the corner of each eye until, with forceps, you can roll back the upper eyelid. The glistening undersurface is a layer of mucous membrane continued on the eyeball, called the *conjunctiva*. At the nasal (inner) corner of each eye is a fold of the conjunctiva, the *nictitating membrane*. (See Fig. 81, p. 136 showing the position of the nictitating membrane in the frog.) Also, at each nasal corner is the opening to the *lacrimal ducts*, which receive "tears" from the *lacrimal glands*. Wash the eyes and drain the fluid through a *nasolacrimal duct*. You may be able to see the *Mei-*

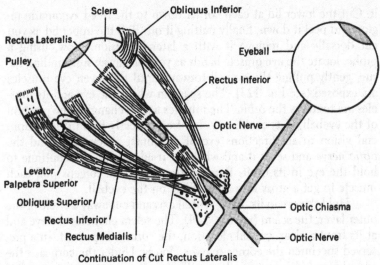

Figure 123. Muscles of Eye—Schematic.

bomian glands, oil glands located on the internal surface of the eyelids. Look for thin yellow bands.

Cut along the dorsal margin of the end of the maxillary-mandibular joint, the *zygomatic arch,* removing the tissue around it to free

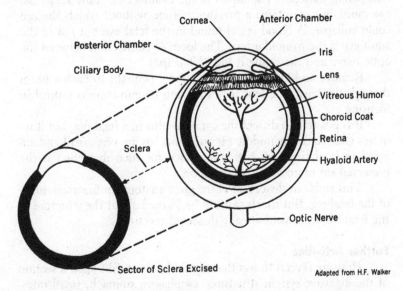

Figure 124. Structure of Fetal Pig Eye—Shallow Cup of Sclera Removed.

203

it. Cut the lower lid at each corner down to the freed zygomatic region and peel it down, finally cutting it off. Pull the upper lid as you cut dorsally and remove it with a lateral incision. Now, using a probe, locate the eye muscle bands as you pick away surrounding tissue, gently pulling the eyeball forward until the seven eye muscles are exposed (see Fig. 123). The space in which the eye and its muscles are found is the *orbit*. The muscles act by changing the position of the eyeball, thus enabling vision from directly forward to peripheral vision in all directions except, of course, backward. Find the *optic nerve* and sever it and whatever residues of tissue continue to hold the eye in its orbit. Remove the eye. With forceps pull each muscle to get a gross idea of its effect on the eyeball.

Hold the eye in its dorso-ventral axis and cut away a circle of the outer layer, the *sclera* (see Fig. 124). The sclera encircles the eye and at its front a clear, curved elevation, the *cornea*, is present. In a preserved specimen the cornea may be cloudy. Under the cornea is the *iris*, pigmented except in albinos. The small opening in the center of the iris is the *pupil*. Under the sclera is a dark layer, the *choroid coat*, and close behind it is the *retina*, richly supplied with nerve endings that connect with the *optic nerve*. Behind the cornea is the *anterior chamber* filled with *aqueous humor*. Behind the iris is a small space, the *posterior chamber*, also filled with liquid. Extending into the posterior chamber is a relatively large *crystalline lens* which also partially occupies the large chamber that is filled with a viscous, gelatinous, transparent material. The liquid in the chamber not only keeps the eye moist but maintains a pressure tension without which the eye could collapse. A blood vessel found in the fetal eye but not in the adult eye is the *hyaloid artery*. The locus of connection between the optic nerve and the retina is the *"blind spot."*

Remove and examine the lens. Use a penlight to flash a beam through the lens. The lens is flexible and its curvature is critical in focusing.

It is possible to dissect the ear apparatus in a fetal pig, but it requires skill, since the middle ear and inner ear are very small and encased in hard tissue. A better specimen for such dissection is the preserved ear of an adult cat or rabbit.

This ends our dissection of the gross anatomy or macroanatomy of the fetal pig. But much can still be learned about the structure of the fetal pig from your already dissected specimen.

Further Activities
Here are several things that should be done. Slit open a section of the digestive system structures: esophagus, stomach, small intestine, and so forth. Wash and examine the internal appearance with a

204

hand lens or a binocular microscope. Check with your biology text-book for an explanation of your findings.

Cut thin slices of muscle, liver, kidney, testis, and other parts. Examine them with a binocular microscope. Refer to a book on histology to learn more about the microscopic structure of parts of the body. Compare prepared slides of cardiac (heart) muscle and involuntary muscle from the large intestine. Examine prepared slides of various glands such as the parotid, pancreas, adrenal, pituitary, and so forth, always referring to histology books. Learn to prepare your own stained slides as well. Full directions are to be found in many books on microscopy and on biological techniques.

We have traveled a long way on the adventurous highway of dissection, from the relatively simple earthworm, whose body resembles a train of similar boxcars, to the complex fetal pig, whose anatomy resembles that of humans. On the way we discovered that every dissection unfolds a story of the past and provides us with a basis for knowing more about the animals of the present. Not only for man is it true that "All the world's a stage." Every animal plays its part in the great drama of life!

Now, it is time to discover that plants too can be dissected.

12

Flowers Can Be Dissected Too

THE GLADIOLUS

MOST PEOPLE THINK OF ANIMALS WHEN DISSECTION IS MENtioned. But plants too are excellent subjects for dissection. Some of the greatest advances in science have resulted from knowledge gained through the dissection of plants. This includes advances in our knowledge of heredity, medicine, evolution, cytology, physiology, and other branches of science.

A plant, like an animal, is made up of organs, tissues, and fundamental units called cells. It engages in almost the same functions as an animal, with two main exceptions—it does not ingest, or swallow, solid food and it does not eject solid wastes. Furthermore, unlike an animal, a green plant can manufacture its own food, using the energy radiated from the sun.

Plants have had their own evolution dating back to immemorial time. In fact there was a time early in global history when it would have been well nigh impossible to distinguish between plants and animals. From these "planimals" have descended the countless strange and wonderful creatures and plants that have helped to fashion the face of the earth. Lichens and similar pioneer plants invade desolate rock, grow in the crevices, and prepare living beachheads for other plants. Their remains provide a network for holding water. Animals move in, following the plants, and over long periods of time the face of the earth is literally changed by the animals and plants. From bare rock to soil to forest is a succession that has happened in many parts of the world.

Of all the parts of a plant the flower has most captivated people with its beauty of color and form. Scientifically, the flower is the key to plant evolution, heredity, and classification. The parts of a flower and their arrangement help us determine the group to which it belongs and help us trace its kinship with other plants. For example,

206

botanists divide flowering plants into two groups—the *monocotyledons* and the *dicotyledons*. The dicotyledons have flowers whose parts grow in multiples of five—five *petals*, five *sepals*, etc. The monocotyledons have flowers with multiples of three—three petals, three sepals, and so on. To which group would a rose, possibly growing in your own garden, with 250 petals and five sepals belong?

DISSECTING THE GLADIOLUS

We are going to dissect a member of the monocotyledons, the gladiolus, described in Fig. 125. The gladiolus is a complete flower with relatively large parts. It is simple to dissect and is a fine representative of one of the two great divisions of flowering plants.

Obtain a stalk or stem with several gladiolus flowers from any flower shop. You will need the instruments used in previous dissections. A microscope and a microscope slide would be helpful. Now prepare to dissect as follows:

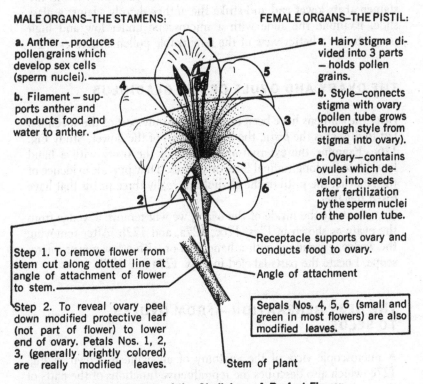

MALE ORGANS—THE STAMENS:

a. Anther —produces pollen grains which develop sex cells (sperm nuclei).

b. Filament — supports anther and conducts food and water to anther.

FEMALE ORGANS—THE PISTIL:

a. Hairy stigma divided into 3 parts — holds pollen grains.

b. Style—connects stigma with ovary (pollen tube grows through style from stigma into ovary).

c. Ovary—contains ovules which develop into seeds after fertilization by the sperm nuclei of the pollen tube.

Receptacle supports ovary and conducts food to ovary.

Angle of attachment

Step 1. To remove flower from stem cut along dotted line at angle of attachment of flower to stem.

Step 2. To reveal ovary peel down modified protective leaf (not part of flower) to lower end of ovary. Petals Nos. 1, 2, 3, (generally brightly colored) are really modified leaves.

Sepals Nos. 4, 5, 6 (small and green in most flowers) are also modified leaves.

Stem of plant

Figure 125. Floral Organs of the Gladiolus—A Perfect Flower.

With scalpel, cut one flower from the stem as shown in Fig. 125. Identify flower parts labeled in Fig. 125.

Carefully peel down modified protective leaf until you reach lower end of ovary. Then cut the leaf off with scissors. Be careful not to damage ovary. With sharp scalpel trim remaining tissue of leaf away from lower end or base of ovary. Bend sepals backward until they crack. Cut them off at point of break with scissors. Remove the petals in the same way. Examine both surfaces of the petals and sepals with a hand lens. Observe how heavily laden they are with sap. Note that the veins show up clearly, indicating that the petal is really a modified leaf with color pigment for attracting insects.

The specimen now looks like a cup with stalks projecting outward. Let's remove each stamen and its attachment from the flower, as shown in Figs. 126 and 126b.

After you have removed the stamens, examine the *anthers* with a low-power microscope or hand lens. Note the enormous number of pollen grains. Can you imagine how many pollen grains are carried from flower to flower by a single bee?

Fig. 126b shows a stamen removed from the flower. Hold the stamen at its lower end and strike the anther sharply against a glass slide. Examine the slide with a microscope under low and high power to get a better view of the remarkable pollen structures.

THE OVARY AND OVULE OF THE GLADIOLUS

After the stamens have been removed, the only part of the specimen left to dissect is the *pistil*, the female organ of the flower. Study Fig. 126c. Examine the external appearance of the ovary with a hand lens. Note the longitudinal lines or sutures. They provide evidence of the fact that the pistil of the gladiolus is really three pistils that have fused together.

To study the inside of the ovary, we will remove a wedge from the ovary as shown in Figs. 126c, 127a, and 127b. After removing the wedge, examine it with a hand lens or with a low-power microscope. Locate the parts labeled in Figs. 127a and 127b.

THE NEXT GENERATION—FROM OVULES
TO SEEDS

A microscopic view of the anatomy of an ovule is shown in Fig. 127c, which also describes the reproductive functions of the parts of the ovule.

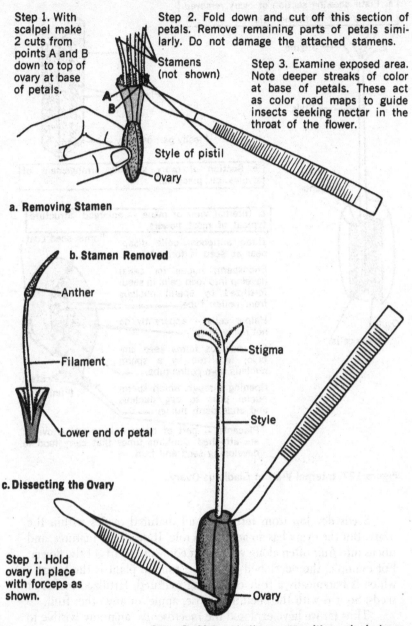

Step 1. With scalpel make 2 cuts from points A and B down to top of ovary at base of petals.

Step 2. Fold down and cut off this section of petals. Remove remaining parts of petals similarly. Do not damage the attached stamens.

Stamens (not shown)

Step 3. Examine exposed area. Note deeper streaks of color at base of petals. These act as color road maps to guide insects seeking nectar in the throat of the flower.

A
B

Style of pistil

Ovary

a. Removing Stamen

b. Stamen Removed

Anther

Filament

Lower end of petal

Stigma

Style

c. Dissecting the Ovary

Step 1. Hold ovary in place with forceps as shown.

Ovary

Step 2. Make shallow cuts with scalpel down two of the division lines. Angle cuts so as to cut out wedge of ovary, as if cutting slice of cake.

Figure 126. Removing Stamens and Dissecting Ovary of the Gladiolus.

209

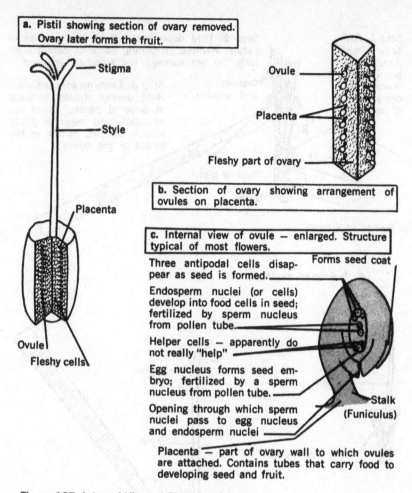

a. Pistil showing section of ovary removed. Ovary later forms the fruit.

— Stigma

—Style

Placenta

Ovule

Fleshy cells

Ovule —

Placenta —

Fleshy part of ovary —

b. Section of ovary showing arrangement of ovules on placenta.

c. Internal view of ovule — enlarged. Structure typical of most flowers.

Three antipodal cells disappear as seed is formed.

Forms seed coat

Endosperm nuclei (or cells) develop into food cells in seed; fertilized by sperm nucleus from pollen tube.

Helper cells — apparently do not really "help"

Egg nucleus forms seed embryo; fertilized by a sperm nucleus from pollen tube.

Opening through which sperm nuclei pass to egg nucleus and endosperm nuclei

Stalk (Funiculus)

Placenta — part of ovary wall to which ovules are attached. Contains tubes that carry food to developing seed and fruit.

Figure 127. Internal View of Gladiolus Ovary.

Seeds develop from fertilized and matured ovules within the ovary. But the ovary has an additional role. It develops, matures, and ripens into *fruit* often along with other connected parts of the flower. For example, the developed ovary of a tomato plant is the tomato, which is botanically a fruit containing matured, fertilized ovules or seeds. So it is with the cucumber, rose, apple, or any other fruit.

Thus far we have explored the *macroscopic* anatomy (visible to unaided eye) of animals and of a flower, with just a glimpse of microscopic anatomy. This opens the door to a long and exciting corridor that leads into many other fields of exploration. For example, after studying the appearance and parts of a heart with the naked eye and with a magnifying lens as we did in Chapter 10, we can go a step further. We can examine the microscopic anatomy or histology of plants and animals, or their parts. Ultimately, as a scientist, you will use the electron microscope, which magnifies to about 200,000 diameters.

Project 1: Studying Microscopic Anatomy

How can you study tissues and cells of organisms? Let's take the ovary of the gladiolus as an example. Figs. 127a and 127b show the inside of an ovary, and Fig. 127c shows the inside of an ovule. To see these structures microscopically you will need a special instrument called a *microtome*. This is used to cut thin slices of the organism or of its parts. The pieces are so thin that they do not block out transmitted light as you examine them under a microscope. This transmitted light penetrates the slide and the transparent parts of the sectioned specimen. You see the opaque parts of the cells because they block out light rays and cast shadow outlines.

There are some simple, inexpensive microtomes on the market that can help you to learn a great deal about tissues. You must learn how to prepare tissues for slicing and for staining, which shows up, in color, the structures you wish to study. These techniques can be learned from a book on histology. The materials are available. The subject is fascinating. What are you waiting for?

Project 2: Science Collections in Plastic (Fig. 128)

It is now possible to preserve your plant and animal specimens by embedding or encasing them in beautiful clear plastic, shaped into any form you like. For example, you can preserve butterflies in

Figure 128. Preserving Specimen by Embedding in Plastic. (Courtesy Carolina Biological Supply Company.)

211

transparent plastic shaped to form bookends, paperweights, or whatever your creative mind suggests. These plastic mounts last indefinitely and remain clear as crystal. You can arrange combinations of animals and plants in these plastic molds to imitate their natural habitat or ecological setting. You can even go into business for yourself and sell beautiful gifts for Mother's Day, Father's Day, and other festive occasions.

Materials and directions for using the plastic encasing chemicals are sold by biological supply houses. Try this technique for pleasure and for beauty. Have fun and profit from it.

Project 3: Survival Problems in Manned Spaceships

Man has already circumnavigated the earth in a spaceflight, though only for a short duration. There is enough evidence to show that we are ready to reach greater distances into space. However, keeping humans alive in prolonged spaceflights is one horn of another dilemma. They must have a continuous supply of oxygen and means of continuously removing carbon dioxide. In addition, adequate food must be available, humidity must be controlled, temperature must be regulated, and the astronauts must be protected against harmful radiations. These are only a few of the difficult problems that have to be solved to ensure survival in extended spaceflight.

All these problems have to be solved at one point, the spaceship, traveling in utter loneliness and in darkness through unlivable space. Would you like to try your hand at one of the complex problems that has yet to be solved? The problem is, how to supply the astronaut with adequate food that, in turn, can also supply adequate oxygen and, moreover, remove from the spaceship the deadly carbon dioxide man exhales—*all at the same time!* Caution: This is no cookbook project. It's wide open, and no one can predict just where it will lead. Well, let's look into the problem a little more closely.

The spaceship has no room for storage of large amounts of food and for oxygen tanks. Therefore a way must be found to have food and oxygen continuously produced within the spaceship or to "hibernate" the spaceman so that he will have no need for food and oxygen. We will try the first alternative, which seems more practicable at the present level of knowledge.

We know that green plants can convert the sun's energy into food with the aid of water, raw materials, and carbon dioxide. However, we can't plant a vegetable garden in a spaceship that will do all of these things. But a microscopic green plant cell belonging to a group of primitive green plants known as *algae* can do these things—*Chlorella vulgaris.* It grows in water which the astronauts

212

will need, and it requires only carbon dioxide, light, temperature control, and some raw materials (chemicals) which are easily supplied, to manufacture food.

Experiments have shown that *Chlorella vulgaris* is an efficient food producer, rich in the nutrients needed by man. But can it be eaten by humans? Yes. *Chlorella vulgaris* has been used to make bread, soups, and even ice cream. Here is one suggestion to find out whether *Chlorella vulgaris* can produce enough oxygen and remove enough carbon dioxide to keep animals alive.

Obtain two 3- or 4-gallon aquarium tanks, each with a glass top. Add about 1 inch of water to the bottom of each tank. Place an equal number of live frogs, from six to twelve, in each tank. They will produce the carbon dioxide needed for the chlorella experiment. Cover each tank with its glass top. Keep the tanks in the light near a window but *not in direct sunlight.*

Prepare a culture medium for the *Chlorella vulgaris.* Both the *Chlorella vulgaris* and directions for preparing the medium may be obtained from biological supply houses. One medium used successfully is a mixture of 0.25 g $MgSO_4 \cdot 7H_2O$, 0.25 g KH_2PO_4, 0.12 g KCl, and 0.01 cc of a 1% solution of $FeCl_3 \cdot 6H_2O$ stirred in 250 ml of water to which 1.0 g of $Ca(NO_3)_2$ has been added.

Pour this mixture into five wide-mouthed 50- to 100-ml bottles. Inoculate chlorella culture into each bottle with a transfer needle. (See your science teacher for explanation of transfer-needle technique.) Cover each bottle with gauze. Place a drop of chlorella culture on a slide with a transfer needle or a medicine dropper and, with the high power of a microscope, estimate the number of cells inoculated into each bottle. Place bottles near window for several days and make a population count daily until you see a greenish tint in the water, especially near the bottom. (Ask your science teacher to explain the population-count technique.) At this point there are enough chlorella cells to produce life-sustaining oxygen.

Place the five bottles into one tank. The other tank will act as a control to the experiment. Smear the top edges of each tank with glycerine and cover each tank with its glass cover. The glycerine will prevent leakage of air. At 3- to 4-hour intervals daily, for 2 to 3 weeks, watch the frogs' behavior in both tanks. Make respiration counts once a day. (For respiration-count technique see page 152.) Tabulate the respiration counts and draw your own conclusions. Watch the control group of frogs carefully because they are likely to weaken due to the diminishing supply of oxygen in the sealed chamber. The control group has no chlorella to produce oxygen and to take up the carbon dioxide wastes produced by the frogs. When the respiration rate of the control group becomes distinctly different from that of the experimental group, remove the glass top and replace it with a piece of wire screening.

At the end of the experiment remove bottles containing chlorella from tank and make a population count.

The alert investigator will want to refine the experiment in many ways. For example, temperature of the "spaceship" is an important factor in survival. You can devise methods of controlling and testing different temperature effects on the growth of chlorella, liberation of oxygen, consumption of carbon dioxide, and on the respiration of the amphibian "spacemen" used in the experiment. Can you explain how this setup imitates the conditions of a manned spaceship? Does this project suggest any additional experiments that will have to be done for manned flight into outer space? Can you make your own improvements to obtain additional data from this project?

This ends our dissections, projects, and experiments. We hope your experiences with this book will be a first step toward the development of a lifelong interest in science and possibly toward a richly rewarding career. The door is open. Science needs you and you need science.

Index

Italic numbers refer to illustrations.

216

Urethra, pig's, 168, 169, 170, 171
Urinary bladder, pig's, 170
Urinary ducts, dogfish shark's, 100
Urinary papilla, dogfish shark's, 100, 105
Urogenital papilla, 114
  pig's, 169, 171
Urogenital pore, pig's, 169, 171
Urogenital sinus, pig's, 168–169, 171
Urogenital system
  dogfish shark, 103, *104, 106*
  fetal pig, 167, *168, 169,* 169–171
  frog, 147, *148,* 149
Uterus
  dogfish shark's, 105
  pig's, 171

Vagina
  grasshopper's, 46
  pig's, 169, 171
Valve, clam, 53
Vas deferens
  grasshopper's, 46

Vas deferens (*cont.*)
  perch's, 128
  squid's, 82
Veins, grasshopper's, 50
Vena cava, frog's, 146
Venous circulation, pig's, 177–184
Ventral, 7
Vertebrae, perch's, 126–127, 128
Vertebrates, 91
  general structure of, 159, *160, 161, 162*
Viscera, perch's, 115

Water circulation, clam's, 58
Wings, grasshopper's, 39
Wolffian duct, dogfish shark's, 103
Worm. *See* Earthworm.

Yerkes Maze, 22, *23*

Zygomatic arch, pig's, 203
Zygotes
  clam's, 69, 70
  perch's, 128

223

INCHES

CENTIMETERS

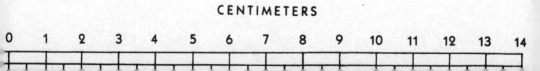